Lecture Notes in Electrical Engineering

Volume 266

For further volumes:
http://www.springer.com/series/7818

About this Series

"Lecture Notes in Electrical Engineering (LNEE)" is a book series which reports the latest research and developments in Electrical Engineering, namely:

- Communication, Networks, and Information Theory
- Computer Engineering
- Signal, Image, Speech and Information Processing
- Circuits and Systems
- Bioengineering

LNEE publishes authored monographs and contributed volumes which present cutting edge research information as well as new perspectives on classical fields, while maintaining Springer's high standards of academic excellence. Also considered for publication are lecture materials, proceedings, and other related materials of exceptionally high quality and interest. The subject matter should be original and timely, reporting the latest research and developments in all areas of electrical engineering.

The audience for the books in LNEE consists of advanced level students, researchers, and industry professionals working at the forefront of their fields. Much like Springer's other Lecture Notes series, LNEE will be distributed through Springer's print and electronic publishing channels.

Marcin Witczak

Fault Diagnosis and Fault-Tolerant Control Strategies for Non-Linear Systems

Analytical and Soft Computing Approaches

 Springer

Marcin Witczak
Institute of Control and Computation
  Engineering
University of Zielona Góra
Zielona Góra
Poland

ISSN 1876-1100   ISSN 1876-1119   (electronic)
ISBN 978-3-319-03013-5   ISBN 978-3-319-03014-2   (eBook)
DOI 10.1007/978-3-319-03014-2
Springer Cham Heidelberg New York Dordrecht London

Library of Congress Control Number: 2013951773

Printed on acid-free paper

Springer is part of Springer Science+Business Media (www.springer.com)

To my beloved and wonderful wife Anna

Preface

A permanent increase in the complexity, efficiency and reliability of modern industrial systems necessitates a continuous development in control and fault diagnosis. A moderate combination of these two paradigms is intensively studied under the name of fault-tolerant control. This real world's development pressure has transformed fault diagnosis and fault-tolerant control, initially perceived as the art of designing a satisfactorily safe system, into the modern science that it is today.

Indeed, the classic way of fault diagnosis boils down to controlling the limits of single variables and then using the resulting knowledge for fault alarm purposes. Apart from the simplicity of such an approach, the observed increasing complexity of modern systems necessitates the development of new fault diagnosis techniques. On the other hand, the resulting fault diagnosis system should be suitably integrated with the existing control system in order to prevent the development of faults into failures, perceived as a complete breakdown of the system being controlled and diagnosed.

Such a development can only be realised by taking into account the information hidden in all measurements. One way to tackle such a challenging problem is to use the so-called model-based approach. Indeed, the application of an adequate model of the system being supervised is very profitable with respect to gaining the knowledge regarding its behaviour. A further and deeper understanding of the current system behaviour can be achieved by implementing parameter and state estimation strategies. The obtained estimates can then be used for supporting diagnostic decisions and increasing the control quality, while the resulting models (along with the knowledge about their uncertainty) can be used for designing suitable control strategies.

Although the majority of industrial systems are nonlinear in their nature, the most common approach to settle fault diagnosis and fault-tolerant control problems is to use well-known tools for linear systems, which are widely described and well documented in many excellent monographs and books. On the other hand, publications on integrated fault diagnosis and fault-tolerant control for nonlinear systems are scattered over many papers and a number of book chapters.

Taking into account the above-mentioned conditions, this book presents selected *Fault Diagnosis and Fault-Tolerant Control Strategies for Non-Linear Systems* in a unified framework. In particular, starting from advanced state estimation strategies up to modern soft computing, the discrete-time description of the system is employed. Such a choice is dictated by the fact that the discrete-time description is easier and more natural to implement on modern computers than its continuous-time counterpart. This is especially important for practicing engineers, who are hardly ever fluent in complex mathematical descriptions.

The book results from my research in the area of fault diagnosis and fault-tolerant control for nonlinear systems that has been conducted since 1998. It is organised as follows. Part I presents original research results regarding state estimation and neural networks for *Robust Fault Diagnosis*. Part II is devoted to the presentation of integrated fault diagnosis and fault-tolerant systems. It starts with a general fault-tolerant control framework, which is then extended by introducing robustness with respect to various uncertainties. Finally, it is shown how to implement the proposed framework for fuzzy systems described by the well-known Takagi–Sugeno models.

This book is primarily a research monograph which presents, in a unified framework, some recent results on fault diagnosis and fault-tolerant control of nonlinear systems. It is intended for researchers, engineers and advanced post-graduate students in control and electrical engineering, computer science, as well as mechanical and chemical engineering.

Some of the research results presented in this book were developed with the kind support of the National Science Centre in Poland under the grant No. NN514678440 on *Predictive fault-tolerant control for nonlinear systems*.

I would like to express my sincere gratitude to my family for their support and patience. I am also grateful to Prof. Józef Korbicz for suggesting the problem, and for his continuous help and support. I would like to express my thanks to my friends Prof. Vicenç Puig (Universidad Politécnica de Cataluña) and Prof. Christophe Aubrun (Université de Lorraine) for the long lasting and successful cooperation. I also would like to express my special thanks to Dr. Lukasz Dziekan for his help in preparing some of the computer programmes, laboratory experiments and simulations of Chap. 6

Zielona Góra, August 2013   Marcin Witczak

Contents

Symbols

t	Time
k	Discrete time
$\varepsilon(\cdot)$	Expectation operator
$\mathrm{Co}(\cdot, \ldots, \cdot)$	Convex hull
$\mathrm{col}(A)$	Column space of A
$\mathrm{hull}(\cdot)$	Convex hull
$\underline{\sigma}(\cdot), \overline{\sigma}(\cdot)$	Minimum, maximum singular value
$\mathcal{N}(\cdot, \cdot)$	Normal distribution
$\mathcal{U}(\cdot, \cdot)$	Uniform distribution
$A \succ (\succeq)\mathbf{0}$	Positive (semi-)definite matrix
$A \prec (\preceq)\mathbf{0}$	Negative (semi-)definite matrix
$\rho(A)$	Spectral radius of A
$x_k, x_{f,k}, \hat{x}_k, \hat{x}_{f,k} \in \mathbb{R}^n$	State vector and its estimate
$y_k, y_{f,k}, \hat{y}_k, \hat{y}_{f,k} \in \mathbb{R}^m$	Output vector and its estimate
$y_{M,k} \in \mathbb{R}^m$	Model output
$e_k \in \mathbb{R}^n$	Estimation or tracking error
$\varepsilon_k \in \mathbb{R}^m$	Output error
$u_k, u_{f,k} \in \mathbb{R}^r$	Input vector
$d_k \in \mathbb{R}^q$	Unknown input vector, $q \leq m$
$z_k \in \mathbb{R}^m$	Residual
$r_k \in \mathbb{R}^{n_p}$	Regressor vector
w_k, v_k	Process and measurement noise
Q_k, R_k	Covariance matrices of w_k and v_k
$p \in \mathbb{R}^{n_p}$	Parameter vector
$f_k \in \mathbb{R}^s$	Fault vector
$g(\cdot), h(\cdot)$	Non-Linear functions
$E_k \in \mathbb{R}^{n \times q}$	Unknown input distribution matrix
$L_k, L_{a,k}, L_{s,k}$	Fault distribution matrices
n_t, n_v	Number of measurements
n_h	Number of hidden neurons

Abbreviations

AFD	Active Fault Diagnosis
AIC	Akaike Information Criterion
ANN	Artificial Neural Network
ARS	Adaptive Random Search
BEA	Bounded-Error Approach
D-OED	D-Optimum Experimental Design
DMVT	Differential Mean Value Theorem
EA	Evolutionary Algorithm
EKF	Extended KF
ESSS	Evolutionary Search with Soft Selection
EUIO	Extended UIO
FDD	Fault Detection and Diagnosis
FDI	Fault Detection and Isolation
FIM	Fisher Information Matrix
FPC	Final Prediction Error
FTC	Fault-Tolerant Control
GA	Genetic Algorithm
GMDH	Group Method of Data Handling
GP	Genetic Programming
ICSP	Interval Constraint Satisfaction Problem
IMM	Interactive Multiple Model
KF	Kalman Filter
LPV	Linear Parameter-Varying
LSM	Least-Square Method
LTI	Linear Time-Invariant
MCE	Monte Carlo Evaluation
MIMO	Multi-Input Multi-Output
MISO	Multi-Input Single-Output
MLP	Multi-Layer Perceptron
OED	Optimum Experimental Design
PFD	Passive Fault Diagnosis
RBF	Radial Basis Function
RLS	Recursive Least-Square
RTD	Resistive Thermal Device

SISO	Single-Output Single-Input
SIVIA	Set Inversion Via Interval Analysis
UIF	Unknown Input Filter
UIO	Unknown Input Observer
UKF	Unscented Kalman Filter
UT	Unscented Transform

Chapter 1
Introduction

A permanent increase in the complexity, efficiency, and reliability of modern industrial systems necessitates a continuous development in control and Fault Diagnosis (FD) [1–6] theory and practice. A moderate combination of these two paradigms is intensively studied under the name Fault-Tolerant Control (FTC) [7–10]. FTC [1] is one of the most important research directions underlying contemporary automatic control. It can also be perceived as an optimised integration of advanced fault diagnosis [5, 6, 11] and control [1] techniques. Engineers have investigated the occurrence and impact of faults for a long time, due to their potential to cause substantial damage to machinery and risk for human health or life. Nowadays, the research and applications of FD and FTC extend beyond the normally accepted safety-critical systems of nuclear reactors, chemical plants or aircrafts, to new systems such as autonomous vehicles or fast rail systems. Early detection and maintenance of faults can help avoid system shutdown, breakdowns and even catastrophes involving human fatalities and material damage. A rough scheme of the modern control system that is able to tackle such a challenging problem is presented in Fig. 1.1 [6, 12].

As can be observed, the controlled system is the main part of the scheme, and it is composed of actuators, process dynamics and sensors. Each of these parts is affected by the so-called unknown inputs, which can be perceived as process and measurement noise as well as external disturbances acting on the system. When model-based control and analytical redundancy-based fault diagnosis are utilised [1, 2, 5–10], then the unknown input can also be extended by model uncertainty, i.e., the mismatch between the model and the system being considered.

The system may also be affected by faults. A fault can generally be defined as an unpermitted deviation of at least one characteristic property or parameter of the system from the normal condition, e.g., a sensor malfunction. All the unexpected variations that tend to degrade the overall performance of a system can also be interpreted as faults. Contrary to the term *failure*, which suggests a complete breakdown of the system, the term *fault* is used to denote a malfunction rather than a catastrophe. Indeed, failure can be defined as a permanent interruption in the system ability to perform a required function under specified operating conditions. This distinction is

M. Witczak, *Fault Diagnosis and Fault-Tolerant Control Strategies for Non-Linear Systems*, 1
Lecture Notes in Electrical Engineering 266, DOI: 10.1007/978-3-319-03014-2_1,
© Springer International Publishing Switzerland 2014

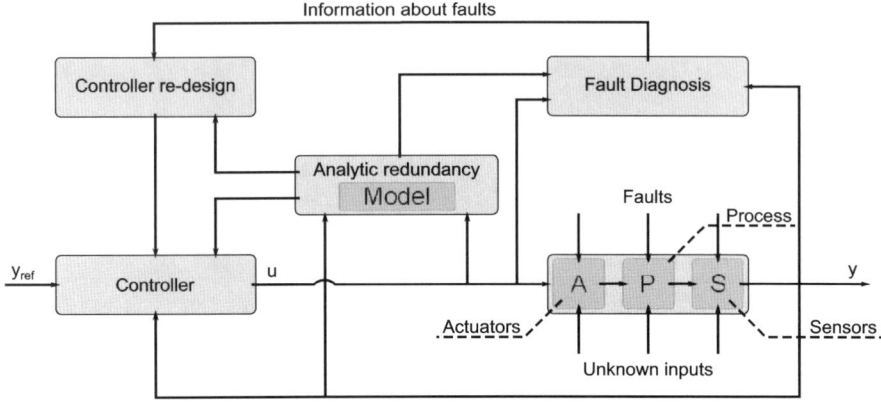

Fig. 1.1 Modern control system

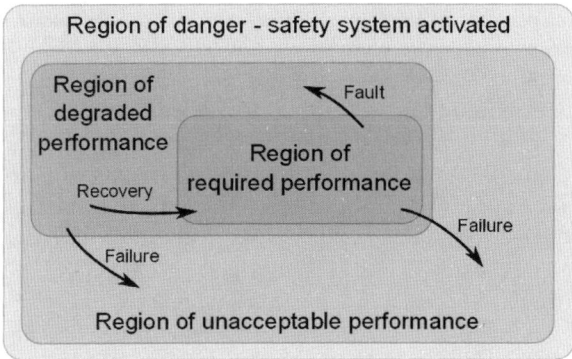

Fig. 1.2 Regions of system performance

clearly illustrated in Fig. 1.2. Since a system can be split into three parts (Fig. 1.1), i.e., actuators, the process, and sensors, such a decomposition leads directly to three classes of faults:

- *Actuator faults*, which can be viewed as any malfunction of the equipment that actuates the system, e.g., a malfunction of the electro-mechanical actuator for a diesel engine [13]. This kind of faults can be divided into three categories:

    Lock-in-place: the actuator is locked in a certain position at an unknown time t_f and does not respond to subsequent commands:

    $$u_{i,k} = u_{i,t_f} = \text{const}, \quad \forall k > t_f. \tag{1.1}$$

    Outage: the actuator produces zero force and moment, i.e., it becomes ineffective:

    $$u_{i,k} = 0, \quad \forall k > t_f. \tag{1.2}$$

Loss of effectiveness: a decrease in the actuator gain that results in a deflection that is smaller than the commanded position:

$$u_{i,k} = k_i u_{i,k}^c, \quad 0 < k_i < 1 \quad \forall k > t_f, \tag{1.3}$$

where $u_{i,k}^c$ stands for the required actuation.

- *Process faults* (or component faults), which occur when some changes in the system make the dynamic relation invalid, e.g., a leak in a tank in a two-tank system.
- *Sensor faults*, which can be viewed as serious measurements variations. Similarly to actuator faults, two sensor fault scenarios can be considered:

Lock-in-place: the sensor is locked in a certain position at an unknown time t_f and does not provide the current value of the measured variable:

$$y_{i,k} = y_{i,t_f} = \text{const}, \quad \forall k > t_f. \tag{1.4}$$

Loss of measurement accuracy: a degradation of the measurement accuracy of the sensor:

$$y_{i,k} = k_i y_{i,k}^c, \quad \forall k > t_f, \tag{1.5}$$

while $y_{i,k}^c$ stands for the true value of the measured variable and k_i is significantly different from 0.

The role of the fault diagnosis part is to monitor the behaviour of the system and to provide all possible information regarding the abnormal functioning of its components. As a result, the overall task of fault diagnosis consists of three subtasks [2] (Fig. 1.3):

Fault detection: to make a decision regarding the system stage–either that something is wrong or that everything works under the normal conditions;
Fault isolation: to determine the location of the fault, e.g., which sensor or actuator is faulty;
Fault identification: to determine the size and type or nature of the fault.

However, from the practical viewpoint, to pursue a complete fault diagnosis the following three steps have to be realised [14]:

Residual generation: generation of the signals that reflect the fault. Typically, the residual is defined as a difference between the outputs of the system and its estimate obtained with the mathematical model;
Residual evaluation: logical decision making on the time of occurrence and the location of faults;
Fault identification: determination of the type of fault, its size and cause.

The knowledge resulting from these steps is then provided to the controller re-design part, which is responsible for changing the control law in such a way as to maintain

Fig. 1.3 Three-stage process
of fault diagnosis

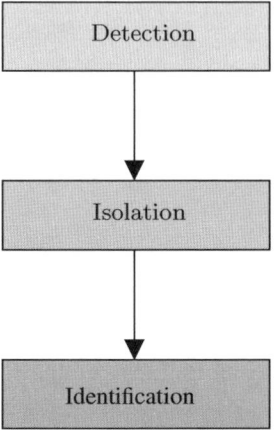

the required system performance. Thus, the scheme presented in Fig. 1.1 can be
perceived as a fault-tolerant one.

Finally, it is worth nothing that the term *symptom* denotes a deviation of an observable quantity from normal behaviour.

Varying nomenclature has been established throughout the years of FD and FTC
development, which can be summarised as follows [1]:

- **Safety** depicts the absence of danger. A safety system is a part of the control equipment with the sole purpose to protecting a technological system from a permanent
  damage (or preventing human casualties). It enables a controlled shutdown, which
  halts the technological process into a safe state. It is capable of evaluating the information about critical signals and enables dedicated actuators to stop the process
  under special conditions. Hence, the overall system is called a *fail-safe system*.
- **Reliability** is the probability that a system performs its intended function for
  a specified time period under normal conditions. Reliability studies the frequency
  with which the system is faulty, but it is not able to provide information about the
  current fault status. FTC cannot change the reliability of the plant components,
  but it can improve the overall reliability because it allows the overall system to
  remain functional even after the occurrence of faults.
- **Availability** is the probability that a system stays operational when needed. As
  opposed to reliability, it is also dependent on the maintenance policies, which are
  applied to the system components.
- **Dependability** accumulates together the three properties of safety, reliability and
  availability. A dependable system is a fail-safe system with high availability and
  reliability.

A fault-tolerant system has the appealing property that faults do not develop into
failures. In the strict form, the system is said to be *fail-operational*, because the
performance remains the same, whereas in a reduced form, the system is operational

after fault occurrence, but with possibly degraded performance. Such a system is then called *fail-graceful*.

The relation between safety and fault tolerance will be elaborated in greater detail because of its importance. Let us assume that the system performance can be described by different regions, as shown in Fig. 1.2.

The system should remain in the region of the required performance during the period of being operational. The controller's aim is to hold the nominal system in this region, in spite of uncertainties and disturbances. It may even hold it during small faults; however, this is not its primary aim. Indeed, the effect of faults is being masked, therefore the fault diagnosis system may not work as required.

On the other hand, the region of degraded performance is the one where the faulty system is allowed to remain. However, the performance of such a system is (in some way) degraded and does not satisfy the nominal performance levels. The system goes from the region of required performance to the degraded one due to faults. The FTC controller should have the capability to perform recovery actions that prevent further degradation of the performance. It is obvious that the optimal strategy is that the system returns to the region of the required performance. At the borderline of these two regions, a supervision system is involved, which diagnoses the faults and reconfigures the controller into the new circumstances.

The region of unacceptable performance should be avoided at all costs, by means of FTC. This region lies between the one of degraded performance and that of danger, which could lead to a disaster. The system goes to this region by either a sudden failure, or due to an uneffective FTC system, i.e., not preventing faults from developing into failures.

To avoid danger for the system and its environment, the safety system stops the operation of the overall system. If the border of the region of unacceptable performance is crossed, the safety system should be immediately involved. This clearly shows that the FTC controller and the safety system work in separate regions of the system performance and fulfil a complementary role. For example, in industrial standards, safety systems and supervision systems are executed as separate units.

1.1 Introductory Background

If residuals are properly generated, then fault detection becomes a relatively easy task. Since without fault detection it is impossible to perform fault isolation and, consequently, fault identification, all efforts regarding the improvement of residual generation seem to be justified. This quality influences also FTC, and hence it deserves a special attention.

There have been many developments in model-based fault detection since the beginning of the 1970s, regarding both the theoretical context and the applicability to real systems (see [2, 5, 15] for a survey). Generally, the most popular approaches can be split into three categories, i.e.,

Fig. 1.4 Simple residual generation scheme

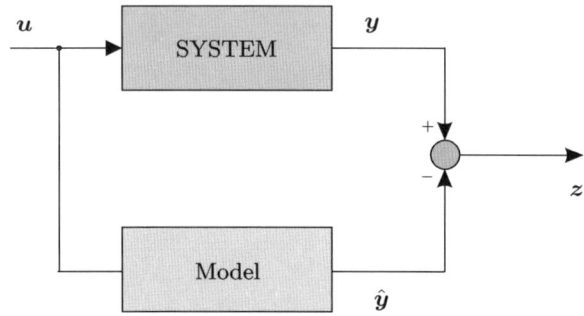

- parameter estimation,
- parity relation,
- observer-based.

All of them, in one way or another, employ a mathematical system description to generate the residual signal. Except for parameter estimation-based FDI, the residual signal is obtained as a difference between the system output and its estimate obtained with its model, i.e.,

$$z_k = y_k - \hat{y}_k. \tag{1.6}$$

The simplest model-based residual generation scheme can be realised in a way similar to that shown in Fig. 1.4. In this case, the design procedure reduces to system identification, and fault detection boils down to checking the norm of the residual signal $\|z_k\|$. In such a simple residual generation scheme, neural networks seem to be especially popular [5, 6].

Irrespective of the identification metod used, there is always the problem of model uncertainty, i.e., the model-reality mismatch. Thus, the better the model used to represent the system behaviour, the better the chance of improving the reliability and performance in diagnosing faults. This is the main reason why the fault detection scheme shown in Fig. 1.4 is rarely used for maintaining fault diagnosis of high-safety systems. Indeed, disturbances as well as model uncertainty are inevitable in industrial systems, and hence there exists a pressure creating the need for robustness in fault diagnosis systems. This robustness requirement is usually achieved at the fault detection stage, i.e., the problem is to develop residual generators which should be insensitive (as far as possible) to model uncertainty and real disturbances acting on a system while remaining sensitive to faults. In one way or another, all the above-mentioned approaches can realise this requirement for linear systems.

Other problems arise in fault detection of non-linear systems. Indeed, the available non-linear system identification techniques limit the application of fault detection. For example, in the case of observer-based FDI, non-linear state-space models cannot be usually obtained using physical considerations (physical laws governing the system being studied). Such a situation is usually caused by the high complexity of the system being examined. This means that a model which merely approximates system-input behaviour (no physical interpretation of the state vector or parameters) should be employed.

The process of fault isolation requires usually more complex schemes than the one of fault detection. Indeed, this very important task of FDI is typically realised by either the so-called dedicated or generalised schemes [2, 3]. In the case of a dedicated scheme, residual generators are designed in such a way that each residual z_i, $i = 1, \ldots, s$, is sensitive to one fault only while remaining insensitive to others. Apart from a very simple fault isolation logic, which is given by

$$|z_{i,k}| > T_i \quad \Rightarrow f_{i,k} \neq 0, \quad i = 1, \ldots, s, \tag{1.7}$$

where T_i is a predefined threshold, this fault isolation design procedure is usually very restrictive and does not allow achieving additional design objectives such as robustness to model uncertainty. Irrespective of the above difficulties, the dedicated fault isolation strategy is frequently used in neural network-based FDI schemes [5].

On the contrary, residual generators of a generalised scheme are designed in such a way that each residual z_i, $i = 1, \ldots, s$, is sensitive to all but one faults. In this case, the fault detection logic is slightly more complicated, i.e.,

$$\left. \begin{array}{l} |z_{i,k}| \leq T_i \\ |z_{j,k}| > T_j, \ j = 1, \ldots, i-1, i+1, \ldots, s \end{array} \right\} \Rightarrow f_{i,k} \neq 0, \quad i = 1, \ldots, s, \tag{1.8}$$

but it requires a less restrictive design procedure than a dedicated scheme, and hence the remaining design objectives can usually be accomplished.

The existing Fault Detection and Diagnosis (FDD) techniques can be, in a general manner, classified into two categories: data-based (model-free) and model-based techniques; each of these methods can additionally be classified as qualitative and quantitative approaches [16].

In essence, a quantitative model-based FDD approach employs a mathematical model (sometimes called analytical redundancy, in contrast to hardware redundancy) to perform FDD tasks in real-time. The most commonly used techniques are based on state estimation, parameter estimation, parity space and some combination of these methods. Due to the fact that most control schemes are model-based, the majority of fault tolerant controllers are designed based on the mathematical model of the system being analysed.

FDD suitable for FTC can be selected based on the following criteria: its capacity to deal with different types of faults (actuator, process and sensor faults), and to supply quick detection, its isolability and identifiability, ease of integrating it with an FTC scheme, its ability to identify multiple faults, robustness to uncertainties and noise, and computational complexity. The comparison of the existing quantitative model-based approaches can be found in [16]. It should be noted that no single method is capable of satisfying all these goals. Though it can be concluded that multiple-model-based, parameter estimation, simultaneous state and parameter estimation techniques are more appropriate for the framework of active FTC [16].

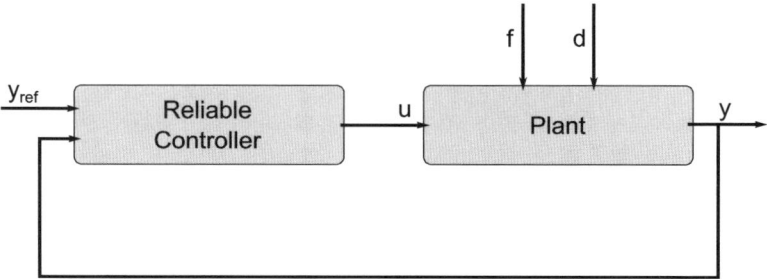

Fig. 1.5 Passive fault-tolerant controller

In general, FTC systems are classified into two distinct classes [16]: passive and active ones. In passive FTC [17–20], controllers are designed to be robust against a set of predefined faults, therefore there is no need for fault detection, but such a design usually degrades the overall performance. Hence, passive FTC sets the control aim in a context where the ability of the system to achieve its given objective is preserved, using the identical control law, irrespective of the system situation (faulty or healthy). Indeed, the control law is not changed when faults occur, so the system is able to achieve its control goal, in general, only for objectives associated with a very low level of performance (sometimes called the *conservative approach*). Further, such a controller works sub-optimally for the nominal plant because its parameters are prearranged so as to get a trade-off between the performance and fault tolerance. It should be noted that a passive fault-tolerant controller is similar to the robust approach when uncertain systems are considered. Although the difference lies not only in the size and interpretation of faults versus uncertainties but also in the structure of the constraints resulting from the faults [1]. An overall structure of passive FTC can be seen in Fig. 1.5. In the literature, passive FTC systems are also known as *reliable* control systems or control systems with *integrity*. However, further discussion of passive FTC is beyond the scope of this book and interested readers are referred to the previously mentioned papers and references therein.

In contrast to passive ones, active FTC schemes react to the system component faults actively by reconfiguring control actions, and by doing so the system stability and acceptable performance is maintained. In certain situations, degraded performance must be accepted. An active FTC (referred from here on simply as FTC, unless some reference to passive FTC must be made) in the literature is sometimes also referred to as self-repairing, reconfigurable, restructurable, or self-designing control systems. To achieve fault tolerance, the control system relies heavily on fault detection and diagnosis to provide the most up-to-date information about the real status of the system [5, 6, 21]. Hence, the main goal of an FTC system is to design the controller with an appropriate architecture, which enables stability and satisfactory performance, not only when all control components are healthy, but also in cases when there are faults in sensors, actuators, or other system components.

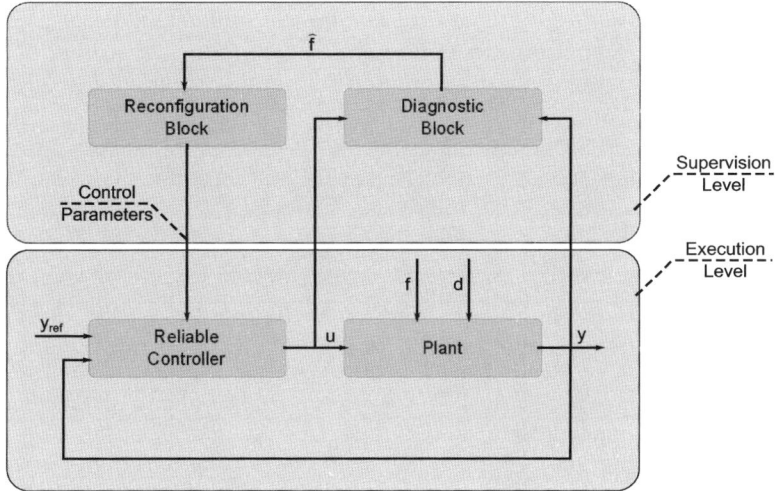

Fig. 1.6 Reconfigurable fault-tolerant controller

The design objectives of the active FTC must include not only transient and tsteady-state performance for the nominal (healthy) system, but additionally for a faulty system. However, the emphasis with respect to system behaviour in both cases is quite different. In healthy conditions, the emphasis is on the performance and overall quality of the system, whereas in the presence of a fault the primary objective is to keep the system from further degradation, even when the nominal performance cannot be achieved (though it should be regained as much as possible).

Usually, as depicted in Fig. 1.6, the FTC system can be divided into four subsystems [16]:

- a reconfigurable controller,
- an FDD scheme,
- a controller reconfiguration scheme,
- a command/reference governor.

It should be noted that the inclusion of both FDD and a reconfigurable controller within the system structure is the main difference between the active and the passive FTC system. Hence, the key issue of a successful FTC scheme is to design a controller which can be easily reconfigured and an FDD scheme that is able to detect faults quickly (being robust to model uncertainties, external disturbances and changing operating conditions). Lastly, a reconfiguration mechanism must be able to recover as much as possible the pre-fault system performance, while working under uncertainties and time-delays intrinsic in FDD as well as the control input and the system state constraints. The key issue in every FTC system is the limited time frame allotted for FDD and reconfiguration of the system controller. Moreover, efficient employment and supervision of available redundancy (in software, hardware and communication networks), while at the same time stabilizing the faulty plant with some performance goals, are some of the main issues to take into account in FTC.

As shown in Fig. 1.6, FDD must provide information about all detected faults in real time. Based on this information, the reconfiguration block must take into consideration the current system state and the outputs, as well as to construct an appropriate post-fault system model. Afterwards, the reconfiguration data for the controller should be designed, in such a way that a currently faulty system is sta-bilised and fault propagation is stopped. The second objective is to recover as much of the nominal performance as possible. Moreover, there is often a need for synthesiz-ing a feed-forward controller in order to guarantee that the closed-loop system tracks a future trajectory during its faulty state. At the same time, the actuator saturation and other system constraints should be taken into consideration and the system trajecto-ries adjusted if necessary. Such an FTC system is often classified as a *reconfigurable* one, though some authors call it an accommodation scheme [1].

However, in some cases reconfiguration of the controller is not enough to stabilize the faulty system. In such cases, the structure of the new controller must be changed. This restructuring also uses an alternative input and output signals in the new con-troller configuration. Afterwards, a new control law has to be designed on-line. Such an FTC controller is called a *restructurable* fault-tolerant controller, and can be seen in Fig. 1.7. This type of FTC is also sometimes called reconfiguration [1], but to avoid confusion the former terms will be used, i.e., reconfigurable versus restructurable. Restructuring of the controller is necessary after occurrence of severe faults that lead to serious changes of the plant behaviour:

- *Actuator failures* interrupt the normal means of controlling the plant and could make the plant partially uncontrollable. Alternative (or redundant) actuators have to be used.
- *Sensor failures* disrupt the information flow between the controller and the plant. They may make the plant partially unobservable. Alternative measurements have to be chosen and used in such a way that the control task is still possible.
- *Plant faults* alter the dynamical behaviour of the overall system. If these alterations cannot be tolerated by any existing control law, then the overall control loop has to be redesigned and a new control law computed.

The necessity of control restructuring is apparent if actuator or sensor failures are contemplated. The total failure of these components leads to a breakdown of the control loop. Hence, a simple adaptation of the controller parameters to a new situation is no longer possible and those having alternative sensors or actuators have to be taken into account, preferably the ones that have similar interactions with the plant and not being under the influence of a fault. Therefore, it is possible to design a controller that satisfies the performance specification of the nominal system [1].

Currently, the existing reconfigurable fault-tolerant control design methods can be classified into one of the following approaches [16, 17]:

- Linear quadratic [22],
- Pseudo-inverse/control mixer [23],

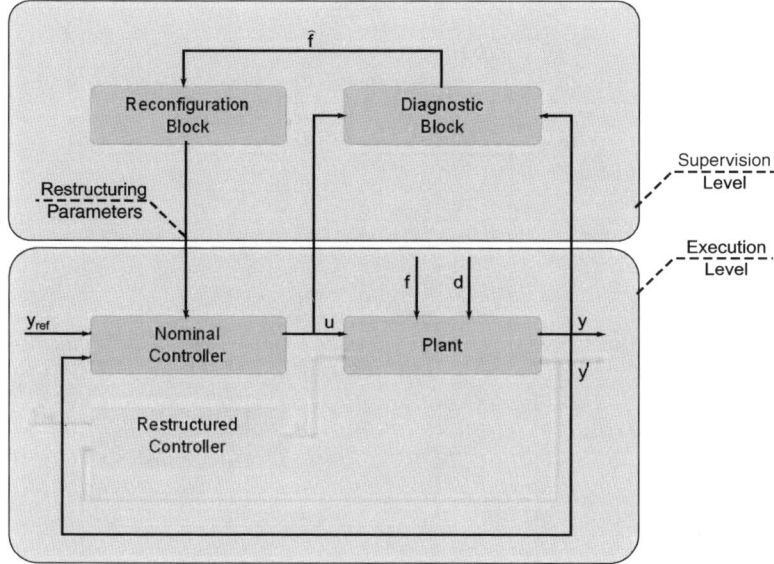

Fig. 1.7 Restructurable fault-tolerant controller

- Intelligent control using expert systems [24], neural networks [25], fuzzy logic [26] and learning methodologies [27],
- Gain scheduling/linear parameter varying [28],
- Adaptive control (model reference) [29],
- Model following [30],
- Multiple-model [31],
- Integrated diagnostics and control [32],
- Eigenstructure assignment [33],
- Feedback linearisation or dynamic inversion [34],
- H_∞ and other robust controls [35],
- Model predictive control [36],
- Linear matrix inequality [37],
- Variable structure and sliding mode control [38],
- Generalised internal model control [39].

FTC methods, as shown in Fig. 1.8, can be also classified in accordance with the following criteria: mathematical design tools, design approaches, reconfiguration mechanisms and the type of systems to be dealt with. The methods (Fig. 1.8) were listed approximately in chronological order to emphasize the historical evolution of FTC design techniques.

In most cases and practical applications, FTC systems rarely use only one of these methods, and to obtain the best possible results a combination of several methods is usually more appropriate. Hence, Fig. 1.8 shows combinations of different control

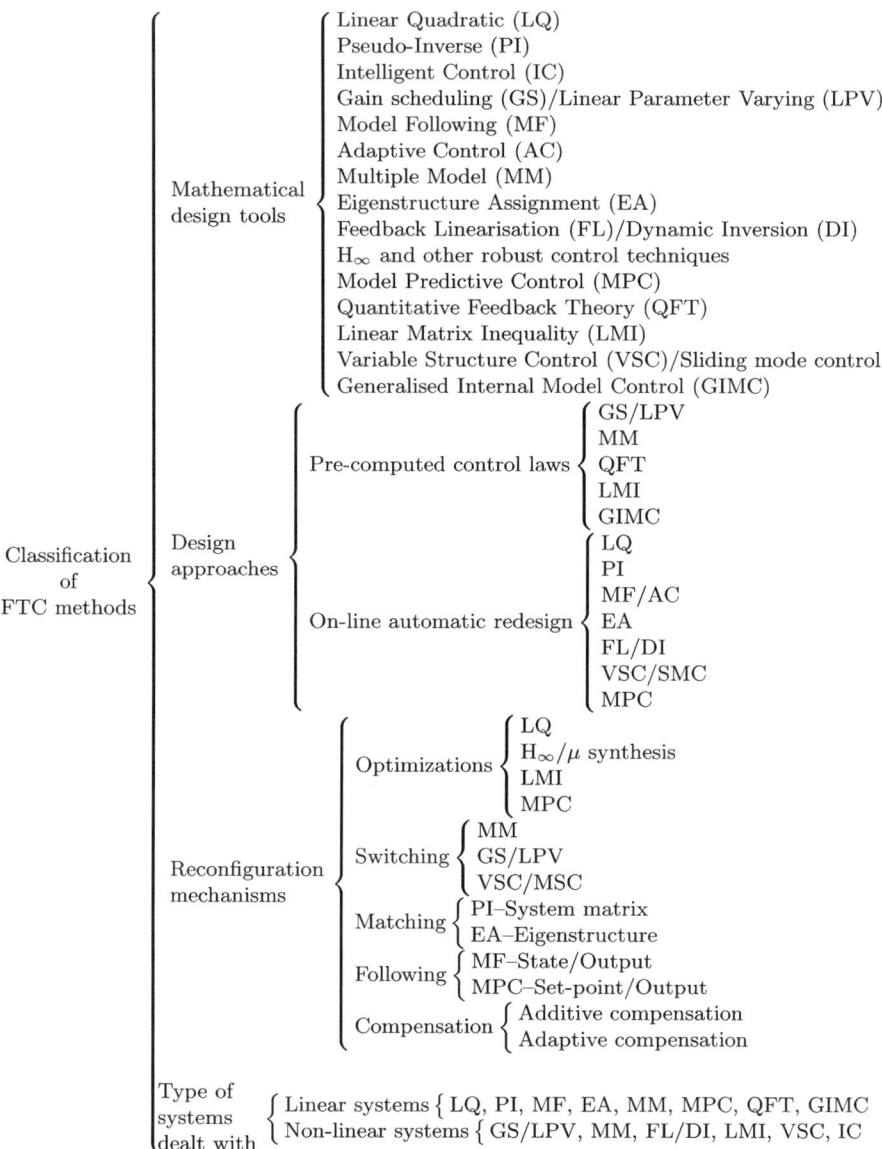

Fig. 1.8 Classification of active FTC systems [16]

structures and control design algorithms frequently used in successful FTC control schemes.

Additionally, many of the currently used FTC design methods rely on ideas originally developed for other control objectives. However, using those well-known control techniques does not mean that new problems and challenges will not appear besides the standard problems found in the conventional controller synthesis.

Finally, in order to judge the adequateness of a control method for FTC, its ability to be implemented in an on-line real-time setting and yet at the same time to be able to maintain acceptable (nominal or degraded) performance is one of the most important criteria. Hence, the following requirements for any technique used in FTC systems can be proposed [16]:

• control reconfiguration must be computed reliably under real-time constraints;
• the reconfigurable controller should be synthesised automatically with as little as possible of trail-and-error and human interactions;
• the selected methods must always provide a solution even if the obtained solution is suboptimal.

As has already been mentioned, a number of books have been published in the last decade on the emerging problem of fault-tolerant control. In particular, the book of [7], which is mainly devoted to fault diagnosis and its applications, provides some general rules for hardware-redundancy based FTC. On the other hand, [8] introduce the concepts of active and passive FTC. The authors also investigate the problem of performance and stability of FTC under imperfect fault diagnosis. In particular, they consider (under a chain of some, not necessarily easy to satisfy, assumptions) the effect of delayed fault detection and imperfect fault identification, but the fault diagnosis scheme is treated separately during the design and no real integration of fault diagnosis and FTC is proposed. FTC is also treated in the very interesting work [9], where a number of practical case studies of FTC are presented, i.e., a winding machine, a three-tank system, and an active suspension system. Unfortunately, in spite of the incontestable appeal of the proposed approaches, neither FTC integrated with fault diagnosis nor a systematic approach to non-linear systems is studied. A particular case of a non-linear aircraft model is studied in [10], but the above-mentioned integration problem is also neglected.

Thus, the main objective of this book is to provide reliable tools for both FD and FTC for non-linear systems along with a suitable integration procedure that will take into account the imprecision of the FD, which impairs the overall performance of FTC.

1.2 Content

The reminder part of this book is divided into two main parts:

Robust fault diagnosis: This part is composed of three chapters. Chapter 2 presents a number of original design strategies for the so-called unknown input observer for both deterministic and stochastic non-linear discrete-time systems. The general idea of unknown input decoupling is shown. As has already been mentioned, the unknown input may represent noise, disturbances as well as model uncertainty. This means that one way to achieve robustness is to decouple the unknown input from the residual. This can be realised with the unknown input observer or filter. Subsequently, careful discussion on unwanted effects of a fault decoupling is given. Indeed, decoupling of the unknown input may result in fault decoupling. This unappealing phenomenon may lead to the so-called effect of undetected faults, which may have serious consequences. Thus, careful analysis regarding such an issue is provided. The rest of the chapter is devoted to different design strategies, which can be applied in various situations. Moreover, a computational framework for deriving the so-called unknown input distribution matrix (shaping the influence of the unknown input on the system) is provided. The remaining chapter is focused on the so-called soft computing techniques [5, 40–44]. Soft computing is an evolving collection of methodologies, which aims to exploit tolerance for imprecision, uncertainty, and partial truth to achieve robustness, tractability, and low cost. It provides an attractive opportunity to represent ambiguity in human thinking with real life uncertainty.

In particular, Chap. 3 is devoted to neural networks, which are one of the main paradigms of soft computing. It also presents original developments regarding neural network-based robust fault diagnosis [6, 12, 45–47]. One objective of this chapter is to show how to describe modelling uncertainty of neural networks. Another objective is to show how to use the resulting knowledge about model uncertainty for robust fault detection. It is also worth noting that this chapter presents experimental design strategies that can be used for decreasing model uncertainty of neural networks [6, 45, 47].

Integrated fault diagnosis and control: This part is composed of three chapters. Chapter 4 introduces the idea of combining fault diagnosis and control schemes within an integrated FTC framework for non-linear systems. In particular, the main idea is to estimate the fault with the unknown input observer by treating the former as an unknown input acting on the system. Subsequently, the information about the fault and an associated fault estimation error is fed into the suitable controller, which aims at compensating the effect of the fault. The originality of the approach follows from the fact that the fault estimation error is taken into account while this unappealing phenomenon is neglected in the approaches presented in the literature [7–10]. The approach presented in Chap. 5 extends the general framework presented in Chap. 4 by relaxing the constraints imposed on system nonlinearities. Moreover, by exploiting the $\mathcal{H}_\infty$ approach, the robustness with respect to the noise/disturbances is achieved. Finally, Chap. 6 shows how to implement the general integrated fault diagnosis and control scheme within the fault-tolerant control framework for non-linear systems described by Takagi–Sugeno models. In particular, the chapter contains a short introduction into fuzzy logic while its remaining part is focused on the so-called Takagi–Sugeno models [5]. A large number of publications both in the control and fault diagnosis frameworks focuses particular attention on this appealing methodology. The chapter contains a short introduction to this computational intelligence method and shows its application to fault diagnosis and fault-tolerant control.

References

1. M. Blanke, M. Kinnaert, J. Lunze, M. Staroswiecki, *Diagnosis and Fault-Tolerant Control* (Springer, New York, 2003)
2. J. Chen, R.J. Patton, *Robust Model Based Fault Diagnosis for Dynamic Systems* (Kluwer Academic Publishers, London, 1999)
3. R. Isermann, *Fault Diagnosis Systems* (An Introduction from Fault Detection to Fault Tolerance (Springer, New York, 2006)
4. J.M. Kościelny, *Diagnostics of Automatic Industrial Processes* (Academic Publishing Office EXIT, Warsaw, 2001)
5. J. Korbicz, J. Kościelny, Z. Kowalczuk, W. Cholewa (eds.), *Fault Diagnosis* (Models, Artificial Intelligence, Applications (Springer, Berlin, 2004)
6. M. Witczak, *Modelling and Estimation Strategies for Fault Diagnosis of Non-linear Systems* (Springer, Berlin, 2007)
7. R. Iserman, *Fault Diagnosis Applications: Model Based Condition Monitoring, Actuators, Drives, Machinery, Plants, Sensors, and Fault-tolerant Systems* (Springer, Berlin, 2011)
8. M. Mahmoud, J. Jiang, Y. Zhang, *Active Fault Tolerant Control Systems: Stochastic Analysis and Synthesis* (Springer, Berlin, 2003)
9. H. Noura, D. Theilliol, J. Ponsart, A. Chamseddine, *Fault-tolerant Control Systems: Design and Practical Applications* (Springer, Berlin, 2003)
10. G. Ducard, *Fault-tolerant Flight Control and Guidance Systems: Practical Methods for Small Unmanned Aerial Vehicles* (Springer, Berlin, 2009)
11. J. Korbicz, J. Kościelny (eds.), *Modeling, Diagnostics and Process Control: Implementation in the DiaSter System* (Springer, Berlin, 2010)
12. M. Witczak, Advances in model-based fault diagnosis with evolutionary algorithms and neural networks. Int. J. Appl. Math. Comput. Sci. **16**(1), 85–99 (2006)
13. M. Blanke, S. Bogh, R. B. Jorgensen, R. J. Patton, Fault detection for diesel engine actuator–a benchmark for FDI, in Proceedings of the 2nd IFAC Symposium on Fault Detection, Supervision and Safety of Technical Processes, SAFEPROCESS'94, vol. 2 (Espoo, 1994), pp. 498–506.
14. P.M. Frank, S.X. Ding, Survey of robust residual generation and evaluation methods in observer-based fault detection systems. J. Process Control **7**(6), 403–424 (1997)
15. R.J. Patton, P.M. Frank, R.N. Clark, *Issues of Fault Diagnosis for Dynamic Systems* (Springer, Berlin, 2000)
16. Y. Zhang, J. Jiang, Bibliographical review on reconfigurable fault-tolerant control systems. Ann. Rev. Control **32**(2), 229–252 (2008)
17. Y. Zhang, J. Jiang, Bibliographical review on reconfigurable fault-tolerant control systems, IFAC Symposium Fault Detection Supervision and Safety of Technical Processes, SAFEPROCESS, (Washington, 2003), pp. 265–276.
18. Y. Liang, D. Liaw, T. Lee, Reliable control of nonlinear systems. IEEE Trans. Autom. Control **45**(4), 706–710 (2000)
19. F. Liao, J. Wang, G. Yang, Reliable robust flight tracking control: an LMI approach. IEEE Trans. Control Syst. Technol. **10**(1), 76–89 (2000)
20. Z. Qu, C.M. Ihlefeld, J. Yufang, A. Saengdeejing, Robust fault-tolerant self-recovering control of nonlinear uncertain systems. Automatica **39**(10), 1763–1771 (2003)
21. H. Li, Q. Zhao, Z. Yang, Reliability modeling of fault tolerant control systems. Int. J. Appl. Math. Comput. Sci. **17**(4), 491–504 (2007)

22. G.H. Yang, J.L. Wang, Y.C. Soh, Reliable LQG control with sensor failures. IEE Proc. Control Theor. Appl. **147**(4), 433–439 (2000)
23. G. Bajpai, B.C. Chang, A. Lau, Reconfiguration of flight control systems for actuator failures. IEEE Aerosp. Electron. Syst. Mag. **16**(9), 29–33 (2001)
24. W. Liu, An on-line expert system-based fault-tolerant control system. Expert Syst. Appl. **11**(1), 59–64 (1996)
25. L.W. Ho, G.G. Yen, Reconfigurable control system design for fault diagnosis and accommodation. Int. J. Neural Syst. **12**(6), 497–520 (2002)
26. A. Ichtev, Multiple fuzzy models for fault tolerant control. Int. J. Fuzzy Syst. **5**(1), 31–40 (2003)
27. Y. Diao, K.M. Passino, Intelligent fault-tolerant control using adaptive and learning methods. Control Eng. Pract. **10**(8), 801–817 (2002)
28. J.Y. Shin, N.E. Wu, C. Belcastro, Adaptive linear parameter varying control synthesis for actuator. J. Guidance Control Dyn. **27**(4), 787–794 (2004)
29. K.-S. Kim, K.-J. Lee, Y. Kim, Reconfigurable flight control system design using direct adaptive method. J. Guidance Control Dyn. **26**(4), 543–550 (2003)
30. Y. Zhang, J. Jiang, Active fault-tolerant control system against partial actuator failures. IEE Proc. Control Theor. Appl. **149**, 95–104 (2002)
31. G.G. Yen, L.W. Ho, Online multiple-model-based fault diagnosis and accommodation. IEEE Trans. Industr. Electron. **50**(2), 296–312 (2003)
32. Y. Zhang, J. Jiang, Integrated active fault-tolerant control using IMM approach. IEEE Trans. Aerosp. Electron. Syst. **37**(4), 1221–1235 (2001)
33. I.K. Konstantopoulos, P.J. Antsaklis, An optimization approach to control reconfiguration. Dyn. Control **9**(3), 255–270 (1999)
34. D.B. Doman, A.D. Ngo, Dynamic inversion-based adaptive/ reconfigurable control of the X-33 on ascent. J. Guidance Control Dyn. **25**(2), 275–284 (2002)
35. G.-H. Yang, J.L. Wang, Y.C. Soh, Reliable H_∞ controller design for linear systems. Automatica **37**(5), 717–725 (2001)
36. M.M. Kale, A.J. Chipperfield, Stabilized MPC formulations for robust reconfigurable flight control. Control Eng. Pract. **13**(6), 771–788 (2005)
37. S. Ganguli, A. Marcos, G. Balas, Reconfigurable LPV control design for boeing 747–100/200 longitudinal axis, in American Control Conference (Anchorage, 2002), pp. 3612–3617.
38. R.A. Hess, S.R. Wells, Sliding mode control applied to reconfigurable flight control design. J. Guidance Control Dyn. **26**(3), 452–462 (2003)
39. D.U. Campos-Delgado, K. Zhou, Reconfigurable fault-tolerant control using GIMC structure. IEEE Trans. Autom. Control **48**(5), 832–839 (2003)
40. R.J. Patton, J. Korbicz, M. Witczak, F.J. Uppal, Combined computational intelligence and analytical methods in fault diagnosis, in *Intelligent Control Systems using Computational Intelligence Techniques*, ed. by A.E. Ruano (The IEE Press, London, 2005), pp. 349–385
41. A.E. Ruano (ed.), *Intelligent Control Systems using Computational Intelligence Techniques* (IEE Press, London, 2005)
42. D. Rutkowska, *Neuro-Fuzzy Architectures and Hybrid Learning* (Springer, Berlin, 2002)
43. L. Rutkowski, *New Soft Computing Techniques for System Modelling, Patern Classification and Image Processing* (Springer, Berlin, 2004)
44. R. Tadeusiewicz, M.R. Ogiela, *Medical Image Understanding Technology* (Artificial Intelligence and Soft Computing for Image Understanding (Springer, Berlin, 2004)
45. M. Witczak, Toward the training of feed-forward neural networks with the D-optimum input sequence. IEEE Trans. Neural Netw. **17**(2), 357–373 (2006)
46. M. Witczak, J. Korbicz, M. Mrugalski, R.J. Patton, A GMDH neural network-based approach to robust fault diagnosis: application to the DAMADICS benchmark problem. Control Eng. Pract. **14**(6), 671–683 (2006)
47. M. Witczak, P. Prętki, Designing neural-network-based fault detection systems with D-optimum experimental conditions. Comput. Assist. Mech. Eng. Sci. **12**(2), 279–291 (2005)

Part I
Robust Fault Diagnosis

Chapter 2
Unknown Input Observers and Filters

As can be observed in the literature, observers (or filters in a stochastic framework) are commonly used in both control and fault diagnosis schemes of non-linear systems (see, e.g., [1–6] and the references therein). Undoubtedly, the most common approach is to use robust observers, such as the Unknown Input Observer (UIO) [2, 7], which can tolerate a degree of model uncertainty and hence increase the reliability of fault diagnosis. Although the origins of UIOs can be traced back to the early 1970s (cf. the seminal work of Ref. [8]), the problem of designing such observers is still of paramount importance both from the theoretical and practical viewpoints. A large amount of knowledge on using these techniques for model-based fault diagnosis has been accumulated through the literature for the last three decades (see Ref. [2] and the references therein). A large number of approaches to non-linear fault diagnosis and fault-tolerant control was published during the last two decades. For example, in Ref. [9] the high gain observer for Lipschitz systems was applied for the purpose of fault diagnosis. One of the standard methods of observer design consists in using a non-linear change of coordinates to turn the original system into a linear one (or a pseudo linear one). As indicated in the literature, such approaches can be applied for fault diagnosis and FTC [10, 11]. It should also be noted that when the feasibility condition regarding the non-linear change of coordinates is not matched, then the celebrated Extended Kalman Filter (EKF) can be applied in both stochastic and deterministic context (see, e.g., [2]).

Generally, design problems regarding the UIOs for non-linear systems can be divided into three distinct categories:

- *Non-linear state transformation-based techniques*: Apart from a relatively large class of systems for which they can be applied, even if the non-linear transformation is possible it leads to another non-linear system and hence the observer design problem remains open (see Ref. [7] and the references therein).
- *Linearisation-based techniques*: Such approaches are based on a similar strategy like that for the EKF [1]. In Ref. [2] the author proposed an extended unknown input observer for non-linear systems. He also proved that the proposed observer is convergent under certain conditions.

M. Witczak, *Fault Diagnosis and Fault-Tolerant Control Strategies for Non-Linear Systems*, 19
Lecture Notes in Electrical Engineering 266, DOI: 10.1007/978-3-319-03014-2_2,
© Springer International Publishing Switzerland 2014

- *Observers for particular classes of non-linear systems*: For example UIOs for polynomial and bilinear systems or for Lipschitz systems [2, 12–14].

In the light of the above discussion, it is clear that accurate state estimation is extremely important for fault detection and control applications. However, estimation under noise and unknown inputs is very difficult.

In order to face the above-mentioned challenges, the design problems regarding UIOs (undertaken within the framework of this chapter) are divided into three distinct categories:

1. How to determine the unknown input distribution matrix, which will not decouple the effect of faults from the residual?
2. How to develop a possibly simple and reliable design procedure of UIO for both non-linear stochastic and deterministic systems?
3. How to extend the approach developed for the constant unknown input distribution matrix into a set of predefined unknown input distribution matrices?

Concerning the first question, a partial answer can be found in Ref. [15]. Indeed, the authors concentrate on the determination of the unknown input distribution matrix for linear systems but they do not answer the question when this matrix will cause the fault decoupling effect. Apart from the fact that there are approaches that can be used for designing UIOs for non-linear systems (listed above), the problem of determining the unknown input distribution matrix for this class of systems remains untouched. In other words, the authors assume that this matrix is known, which apart from relatively simple cases is never the truth. It should also be mentioned that it is usually assumed that disturbance decoupling will not cause a decrease in fault diagnosis sensitivity or fault decoupling in the worst scenario. To tackle this problem within the framework of this chapter, a numerical optimisation-based approach is proposed that can be used to estimate the unknown input distribution matrix which does not cause the fault decoupling effect. As an answer to the second question, this work presents an alternative Unknown Input Filter (UIF) for non-linear systems, which is based on the general idea of the Unscented Kalman Filter (UKF) [16, 17]. This approach is based on an idea similar to that proposed in Refs. [2, 18], but the structure of the scheme is different and instead of the EKF the UKF is employed. To tackle the third problem, it is shown that the Interacting Multiple Model (IMM) algorithm can be employed for selecting an appropriate unknown input distribution matrix from a predefined set. The proposed solutions can be perceived as an alternative to the Takagi–Sugeno-based approach presented, e.g., in Ref. [19], which will be the subject of Chap. 6. Finally, it should be mentioned that some of the results portrayed in this chapter were originally presented in Refs. [18, 20, 21].

2.1 Unknown Input Decoupling

Let us consider a non-linear stochastic system given by the following equations:

$$x_{k+1} = g(x_k) + h(u_k) + Ed_k + Lf_k + w_k, \tag{2.1}$$

$$y_{k+1} = Cx_{k+1} + v_{k+1}. \tag{2.2}$$

Note that the unknown input and fault distribution matrices, denoted by E and L, are assumed (for the sake of simplicity) constant in this section. Such an assumption will be relaxed in Sect. 2.6, where a set of predefined matrices will be used instead. Moreover, it should be mentioned that this chapter focuses on faults that can influence the state equation (2.1), such as actuator faults. The case of sensor faults is beyond the scope of this section and will be investigated in the subsequent part of the book.

The main problem is to design a filter which is insensitive to the influence of the unknown input (external disturbances and modeling errors) while being sensitive to faults. The necessary condition for the existence of a solution to the unknown input decoupling problem is as follows:

$$\text{rank}(CE) = \text{rank}(E) = q \tag{2.3}$$

(see Ref. [2] for a comprehensive explanation). If the condition (2.3) is satisfied, then it is possible to calculate $H = (CE)^+ = \left[(CE)^T CE \right]^{-1} (CE)^T$. Thus, by inserting (2.1) into (2.2) and then multiplying (2.2) by H it is straightforward to show that

$$d_k = H \left[y_{k+1} - C \left[g(x_k) + h(u_k) + Lf_k + w_k \right] - v_{k+1} \right]. \tag{2.4}$$

Substituting (2.4) into (2.1) for d_k gives

$$x_{k+1} = \bar{g}(x_k) + \bar{h}(u_k) + \bar{E} y_{k+1} + \bar{L} f_k + \bar{w}_k, \tag{2.5}$$

where

$$\bar{g}(\cdot) = Gg(\cdot), \quad \bar{h}(\cdot) = Gh(\cdot),$$
$$\bar{E} = EH, \quad \bar{w}_k = Gw_k - EHv_{k+1},$$

and

$$G = I - EHC.$$

Consequently, the general observer structure is

$$\hat{x}_{k+1} = \bar{g}(\hat{x}_k) + \bar{h}(\cdot) + \bar{E} y_{k+1} + K(\cdot), \tag{2.6}$$

where $K(\cdot)$ is the state correction term. In order to make further deliberations more general, no particular form of $K(\cdot)$ is assumed in the present and subsequent section.

Let us define a residual as a difference between the output of the system and its estimate:

$$
\begin{aligned}
z_{k+1} &= y_{k+1} - C\hat{x}_{k+1} \\
&= C(\bar{g}(x_k) - \bar{g}(\hat{x}_k) - K(\cdot)) + \bar{f}_k + C\bar{w}_k + v_{k+1},
\end{aligned}
\tag{2.7}
$$

where

$$
\bar{f}_k = C\bar{L}f_k = C\left[I_n - E\left[(CE)^T CE\right]^{-1}(CE)^T C\right]Lf_k.
\tag{2.8}
$$

A natural question arises: Is it possible that the fault will be decoupled from the residual? If so the proposed strategy seems to be useless as it will lead to undetected faults, which may have serious consequences regarding the performance of the system being diagnosed. An answer is provided in the subsequent section.

2.2 Preventing Fault Decoupling

It is usually assumed that a disturbance decoupling will not cause a decrease in fault diagnosis sensitivity or a fault decoupling in the worst scenario. But such an assumption is a rather unpractical tool in serious applications. Thus, to overcome such a challenging problem, the following theorem provides a simple rule for checking if the proposed unknown input observer will not decouple the effect of a fault from the residual. It relates the fault and unknown input distribution matrices denoted by L and E, respectively. Moreover, let us assume that the following rank condition is satisfied:

$$
\mathrm{rank}(CL) = \mathrm{rank}(L) = s.
\tag{2.9}
$$

Theorem 2.1 *The fault f_k will not be decoupled from the residual (2.7) if and only if the matrix*

$$
[CE \; CL]
\tag{2.10}
$$

is a full-rank one.

Proof Let us suppose (theoretically) that $\mathrm{rank}(C\bar{L}) = s$. Then it can be shown that

$$
f_k = (C\bar{L})^+ \bar{f}_k,
\tag{2.11}
$$

which means that there exists a unique relationship between f_k and $\bar{f}_k$ and hence the fault will not be decoupled from the residual. Unfortunately, the subsequent part of the proof shows that this is not always possible to attain. Indeed, (2.8) can be written into an equivalent form

$$\bar{f}_k = \left[I_m - CE \left[(CE)^T CE \right]^{-1} (CE)^T \right] CLf_k. \tag{2.12}$$

Moreover, it can be observed that

$$\left[I_m - CE \left[(CE)^T CE \right]^{-1} (CE)^T \right]^2$$
$$= I_m - CE \left[(CE)^T CE \right]^{-1} (CE)^T, \tag{2.13}$$

which means that $I_m - CE \left[(CE)^T CE \right]^{-1} (CE)^T$ is an idempotent matrix. One of the fundamental properties of an idempotent matrix is that its rank is equal to the trace, i.e.,

$$\text{rank} \left(I_m - CE \left[(CE)^T CE \right]^{-1} (CE)^T \right)$$
$$= \text{trace} \left(I_m - CE \left[(CE)^T CE \right]^{-1} (CE)^T \right)$$
$$= \text{trace} (I_m) - \text{trace} \left(CE \left[(CE)^T CE \right]^{-1} (CE)^T \right)$$
$$= m - \text{trace} \left(\left[(CE)^T CE \right]^{-1} (CE)^T CE \right) = m - q. \tag{2.14}$$

Thus, from (2.9) it is clear that

$$\text{rank} \left(\left[I_m - CE \left[(CE)^T CE \right]^{-1} (CE)^T \right] CL \right)$$
$$\leq \min(m - q, s). \tag{2.15}$$

On the other hand,

$$\text{rank} \left(\left[I_m - CE \left[(CE)^T CE \right]^{-1} (CE)^T \right] CL \right)$$
$$\geq \text{rank} \left(I_m - CE \left[(CE)^T CE \right]^{-1} (CE)^T \right) + \text{rank} (CL) - m$$
$$= s - q. \tag{2.16}$$

Finally,

$$\max(s - q, 0) \leq \text{rank} \left(\left[I_m - CE \left[(CE)^T CE \right]^{-1} (CE)^T \right] CL \right)$$
$$\leq \min(m - q, s). \tag{2.17}$$

Thus, it is necessary to find an alternative condition under which

$$\begin{aligned}
\bar{f}_k &= CLf_k - CE\left[(CE)^T CE\right]^{-1}(CE)^T CLf_k \\
&= CLf_k - CLf_k = 0.
\end{aligned} \tag{2.18}$$

Indeed, any vector $CLf_k \in \text{col}(CE)$, where

$$\text{col}(CE) = \left\{\alpha \in \mathbb{R}^m : \alpha = CE\beta \text{ for some } \beta \in \mathbb{R}^q\right\}, \tag{2.19}$$

can be written as

$$CLf_k = CE\tilde{f}_k, \tag{2.20}$$

for some non-zero vector $\tilde{f}_k$. As a consequence,

$$\begin{aligned}
& CE\left[(CE)^T CE\right]^{-1}(CE)^T CLf_k \\
&= CE\left[(CE)^T CE\right]^{-1}(CE)^T CE\tilde{f}_k = CE\tilde{f}_k = CLf_k. \tag{2.21}
\end{aligned}$$

From the above discussion, it is clear that the proposed unknown input observer will not decouple the fault effect from the residual iff $CLf_k \notin \text{col}(CE)$, which is equivalent to

$$\text{rank}\left(\left[CE\ CLf_k\right]\right) = q + 1 \tag{2.22}$$

for all $f_{i,k} \neq 0$, $i = 1, \ldots, s$. It is clear that (2.22) is equivalent to the fact that the only solution to (for all $f_{i,k} \neq 0$, $i = 1, \ldots, s$)

$$\alpha_1 (CE)_1 + \alpha_2 (CE)_2 + \cdots + \alpha_q (CE)_q + \alpha_{q+1} CLf_k = 0, \tag{2.23}$$

is for $\alpha_i = 0$, $i = 1, \ldots, q + 1$. By further expansion of (2.23) to

$$\alpha_1 (CE)_1 + \cdots + \alpha_q (CE)_q + \alpha_{q+1} f_{1,k}(CL)_1 + \cdots + \alpha_{q+1} f_{s,k}(CL)_s = 0, \tag{2.24}$$

it can be seen that the zero-valued solution to (2.24) is equivalent to the existence of a full-rank matrix (2.10), which completes the proof.

Since the fault decoupling prevention problem is solved, then it is possible to provide a set of approaches for designing unknown input observers and filters.

2.3 First- and Second-order Extended Unknown Input Observers

The approach presented in this section is dedicated for deterministic systems. The proposed strategy is based on the general framework of the second-order EKF. In particular, the section will show the design of both first- and second-order EUIO. Moreover, to make the presentation more general, the unknown input distribution matrix is assumed to be a time-varying one. Let us consider a non-linear discrete-time system described by (the fault free-case will be considered for the convergence analysis purposes)

$$x_{k+1} = g(x_k) + h(u_k) + E_k d_k, \tag{2.25}$$

$$y_{k+1} = C_{k+1} x_{k+1}. \tag{2.26}$$

The problem is to design an observer that is insensitive to the influence of an unknown input. The necessary condition for the existence of a solution to the unknown input decoupling problem is

$$\text{rank}(C_{k+1} E_k) = \text{rank}(E_k) = q \tag{2.27}$$

(see [15, p. 72, Lemma 3.1] for a comprehensive explanation). If the condition (2.27) is satisfied, then it is possible to calculate $H_{k+1} = (C_{k+1} E_k)^+$, where $(\cdot)^+$ stands for the pseudo-inverse of its argument. Thus, by multiplying (2.26) by H_{k+1} and then inserting (2.25), it is straightforward to show that

$$d_k = H_{k+1} \left[y_{k+1} - C_{k+1} \left[g(x_k) + h(u_k) \right] \right]. \tag{2.28}$$

Substituting (2.28) into (2.25) gives

$$x_{k+1} = \bar{g}(x_k) + \bar{h}(u_k) + \bar{E}_k y_{k+1}, \tag{2.29}$$

where

$$\bar{g}(\cdot) = \bar{G}_k g(\cdot), \quad \bar{h}(\cdot) = \bar{G}_k h(\cdot)$$

$$\bar{G}_k = I - E_k H_{k+1} C_{k+1}, \quad \bar{E}_k = E_k H_{k+1}. \tag{2.30}$$

Thus, the unknown input observer for (2.25) and (2.26) is given as follows:

$$\hat{x}_{k+1} = \hat{x}_{k+1/k} + K_{k+1}(y_{k+1} - C_{k+1}\hat{x}_{k+1/k}),$$

where

$$\hat{x}_{k+1/k} = \bar{g}(\hat{x}_k) + \bar{h}(u_k) + \bar{E}_k y_{k+1}. \tag{2.31}$$

As a consequence, the second-order extended Kalman filter algorithm used for the state estimation of (2.25) and (2.26) can be given as follows:

$$\hat{x}_{k+1/k} = \bar{g}\left(\hat{x}_k\right) + \bar{h}\left(u_k\right) + \bar{E}_k y_{k+1} + s_k, \tag{2.32}$$

$$P_{k+1/k} = \bar{A}_k P_k \bar{A}_k^T + Q_k, \tag{2.33}$$

$$K_{k+1} = P_{k+1/k} C_{k+1}^T$$
$$\cdot \left(C_{k+1} P_{k+1/k} C_{k+1}^T + R_{k+1}\right)^{-1}, \tag{2.34}$$

$$\hat{x}_{k+1} = \hat{x}_{k+1/k} + K_{k+1}(y_{k+1} - C_{k+1}\hat{x}_{k+1/k}), \tag{2.35}$$

$$P_{k+1} = \left[I - K_{k+1} C_{k+1}\right] P_{k+1/k}, \tag{2.36}$$

where

$$\bar{A}_k = \frac{\partial \bar{g}\left(x_k\right)}{\partial x_k}\bigg|_{x_k = \hat{x}_k} = \bar{G}_k \frac{\partial g\left(x_k\right)}{\partial x_k}\bigg|_{x_k = \hat{x}_k}$$
$$= \bar{G}_k A_k, \tag{2.37}$$

and

$$s_{i,k} = \frac{1}{2}\text{trace}\left[P_k \frac{\partial \bar{g}_i\left(x_k\right)^2}{\partial x_k^2}\bigg|_{x_k = \hat{x}_k}\right], \quad i = 1, \ldots, n. \tag{2.38}$$

The algorithm (2.32)–(2.36) can be perceived as the second-order EKF for non-linear systems with an unknown input. It should also be pointed out that when $s_k = 0$ then the algorithm (2.32)–(2.36) reduces to the first-order EUIO.

2.3.1 Convergence Analysis

An important property is the fact that the proposed algorithm is used for the deterministic systems (2.25) and (2.26), and hence there exists a design freedom regarding matrices Q_k and R_k that can be exploited for increasing the convergence rate of the EUIO. To tackle this challenging problem, the convergence conditions of the EUIO related to the matrices Q_k and R_k are developed and carefully analysed.

Using (2.35), the state estimation error can be given as:

$$e_{k+1} = x_{k+1} - \hat{x}_{k+1} = \left[I - K_{k+1} C_{k+1}\right] e_{k+1/k}, \tag{2.39}$$

where

$$e_{k+1/k} = x_{k+1} - \hat{x}_{k+1/k} \approx \bar{A}_k e_k - s_k. \tag{2.40}$$

Assuming that $e_k \neq 0$ and defining an unknown diagonal matrix:

$$\beta_k = \text{diag}(\beta_{1,k}, \ldots, \beta_{n,k}) \tag{2.41}$$

such that

$$-\beta_k e_k = s_k, \tag{2.42}$$

it is possible to write

$$
\begin{aligned}
e_{k+1/k} &= x_{k+1} - \hat{x}_{k+1/k} \\
&= \alpha_k \left[\bar{A}_k + \beta_k \right] e_k = \alpha_k Z_k e_k,
\end{aligned} \tag{2.43}
$$

where $\alpha_k = \text{diag}(\alpha_{1,k}, \ldots, \alpha_{n,k})$ is an unknown diagonal matrix. Thus, using (2.43), the Eq. (2.39) becomes:

$$e_{k+1} = \left[I - K_{k+1} C_{k+1} \right] \alpha_k Z_k e_k. \tag{2.44}$$

It is clear from (2.43) that α_k represents the lineariztion error. This means that the convergence of the proposed observer is strongly related to the admissible bounds of the diagonal elements of α_k. Thus, the main objective of further deliberations is to show that these bounds can be controlled with the use of the instrumental matrices Q_k and R_k.

First let us start with the convergence conditions, which require the following assumptions:

Assumption 1 Following Ref. [22], it is assumed that the system given by (2.26) and (2.29) is locally uniformly rank observable. This guaranties that (see Ref. [22] and the references therein) that the matrix P_k is bounded, i.e., there exist positive scalars $\bar{\theta} > 0$ and $\underline{\theta} > 0$ such that

$$\underline{\theta} I \preceq P_k^{-1} \preceq \bar{\theta} I. \tag{2.45}$$

Assumption 2 The matrix A_k is uniformly bounded and there exists A_k^{-1}. Moreover, let us define

$$\bar{\alpha}_k = \max_{j=1,\ldots,n} |\alpha_{j,k}|, \quad \underline{\alpha}_k = \min_{j=1,\ldots,n} |\alpha_{j,k}|. \tag{2.46}$$

where $\underline{\sigma}(\cdot)$ and $\bar{\sigma}(\cdot)$ denote the minimum and the maximum singular value of their arguments, respectively.

Theorem 2.2 *If*

$$\bar{\alpha}_k \leq \left[\underline{\alpha}_k^2 \frac{\underline{\sigma}(Z_k)^2 \, \underline{\sigma}(C_{k+1})^2 \, \underline{\sigma}(Z_k P_k Z_k^T + Q_k)}{\bar{\sigma}(C_{k+1} P_{k+1/k} C_{k+1}^T + R_{k+1})} \right.
$$

$$+ \; \frac{(1 - \zeta)\underline{\sigma}\left(Z_k P_k Z_k^T + Q_k\right)}{\bar{\sigma}\left(Z_k\right)^2 \bar{\sigma}\left(P_k\right)} \bigg]^{\frac{1}{2}}, \tag{2.47}$$

where $0 < \zeta < 1$, then the proposed extended unknown input observer is locally asymptotically convergent.

Proof The main objective of further deliberations is to determine conditions for which the sequence $\{V_k\}_{k=1}^{\infty}$, defined by the Lyapunov candidate function

$$V_{k+1} = e_{k+1}^T P_{k+1}^{-1} e_{k+1}, \tag{2.48}$$

is a decreasing one. Substituting (2.44) into (2.48) gives

$$V_{k+1} = e_k^T Z_k^T \alpha_k \left[I - C_{k+1}^T K_{k+1}^T\right] P_{k+1}^{-1}$$
$$\times \left[I - K_{k+1} C_{k+1}\right] \alpha_k Z_k e_k. \tag{2.49}$$

Using (2.36), it can be shown that

$$\left[I - C_{k+1}^T K_{k+1}^T\right] = P_{k+1/k}^{-1} P_{k+1}. \tag{2.50}$$

Inserting (2.34) into $\left[I - K_{k+1} C_{k+1}\right]$ yields

$$\left[I - K_{k+1} C_{k+1}\right] = P_{k+1/k} \left[P_{k+1/k}^{-1} - C_{k+1}^T\right.$$
$$\left. \times \left(C_{k+1} P_{k+1/k} C_{k+1}^T + R_{k+1}\right)^{-1} C_{k+1}\right]. \tag{2.51}$$

Substituting (2.50) and (2.51) into (2.49) gives

$$V_{k+1} = e_k^T Z_k^T \alpha_k \left[P_{k+1/k}^{-1} - C_{k+1}^T\right.$$
$$\left. \times \left(C_{k+1} P_{k+1/k} C_{k+1}^T + R_{k+1}\right)^{-1} C_{k+1}\right] \alpha_k Z_k e_k. \tag{2.52}$$

The sequence $\{V_k\}_{k=1}^{\infty}$ is decreasing when there exists a scalar ζ, $0 < \zeta < 1$, such that

$$V_{k+1} - (1 - \zeta)V_k \leq 0. \tag{2.53}$$

Using (2.48) and (2.52), the inequality (2.53) can be written as

$$e_k^T \left[Z_k^T \alpha_k \left[P_{k+1/k}^{-1} - C_{k+1}^T\right.\right.$$
$$\left.\left. \cdot \left(C_{k+1} P_{k+1/k} C_{k+1}^T + R_{k+1}\right)^{-1} C_{k+1}\right] \alpha_k Z_k\right.$$

$$-(1 - \zeta)\boldsymbol{P}_k^{-1}\Big]\boldsymbol{e}_k \leq 0. \tag{2.54}$$

Using the bounds of the Rayleigh quotient for $\boldsymbol{X} \succeq 0$, i.e., $\underline{\sigma}(\boldsymbol{X}) \leq \frac{\boldsymbol{e}_k^T \boldsymbol{X} \boldsymbol{e}_k}{\boldsymbol{e}_k^T \boldsymbol{e}_k} \leq \bar{\sigma}(\boldsymbol{X})$, the inequality (2.54) can be transformed into the following form:

$$\bar{\sigma}\left(\boldsymbol{Z}_k^T \boldsymbol{\alpha}_k \boldsymbol{P}_{k+1/k}^{-1} \boldsymbol{\alpha}_k \boldsymbol{Z}_k\right)$$
$$-\underline{\sigma}\left(\boldsymbol{Z}_k^T \boldsymbol{\alpha}_k \boldsymbol{C}_{k+1}^T \left(\boldsymbol{C}_{k+1}\boldsymbol{P}_{k+1/k}\boldsymbol{C}_{k+1}^T + \boldsymbol{R}_{k+1}\right)^{-1}\right.$$
$$\times \boldsymbol{C}_{k+1}\boldsymbol{\alpha}_k \boldsymbol{Z}_k\Big) - (1 - \zeta)\underline{\sigma}\left(\boldsymbol{P}_k^{-1}\right) \leq 0. \tag{2.55}$$

It is straightforward to show that

$$\bar{\sigma}\left(\boldsymbol{Z}_k^T \boldsymbol{\alpha}_k \boldsymbol{P}_{k+1/k}^{-1} \boldsymbol{\alpha}_k \boldsymbol{Z}_k\right)$$
$$\leq \bar{\sigma}(\boldsymbol{\alpha}_k)^2 \bar{\sigma}(\boldsymbol{Z}_k)^2 \bar{\sigma}\left(\boldsymbol{P}_{k+1/k}^{-1}\right), \tag{2.56}$$

and

$$\underline{\sigma}\left(\boldsymbol{Z}_k^T \boldsymbol{\alpha}_k \boldsymbol{C}_{k+1}^T \left(\boldsymbol{C}_{k+1}\boldsymbol{P}_{k+1/k}\boldsymbol{C}_{k+1}^T + \boldsymbol{R}_{k+1}\right)^{-1} \times \boldsymbol{C}_{k+1}\boldsymbol{\alpha}_k \boldsymbol{Z}_k\right)$$
$$\geq \underline{\sigma}(\boldsymbol{\alpha}_k)^2 \underline{\sigma}(\boldsymbol{Z}_k)^2 \underline{\sigma}\left(\boldsymbol{C}_{k+1}^-\right)^2$$
$$\times \underline{\sigma}\left(\left(\boldsymbol{C}_{k+1}\boldsymbol{P}_{k+1/k}\boldsymbol{C}_{k+1}^T + \boldsymbol{R}_{k+1}\right)^{-1}\right)$$
$$= \frac{\underline{\sigma}(\boldsymbol{\alpha}_k)^2 \underline{\sigma}(\boldsymbol{Z}_k)^2 \underline{\sigma}\left(\boldsymbol{C}_{k+1}^-\right)^2}{\bar{\sigma}\left(\boldsymbol{C}_{k+1}\boldsymbol{P}_{k+1/k}\boldsymbol{C}_{k+1}^T + \boldsymbol{R}_{k+1}\right)}. \tag{2.57}$$

Applying (2.56) and (2.57) to (2.55) and then using (2.33), one can obtain (2.47).

Thus, if the condition (2.47) is satisfied, then $\{V_k\}_{k=1}^{\infty}$ is a decreasing sequence and hence, under the local uniform rank observability condition [22], the proposed observer is locally asymptotically convergent.

2.3.2 Design Principles

First-order Case

When first-order expansion is employed, then $\boldsymbol{s}_k$ in (2.32) should be $\boldsymbol{s}_k = \boldsymbol{0}$. This means that $\boldsymbol{\beta}_k = \boldsymbol{0}$ and hence $\boldsymbol{Z}_k = \bar{\boldsymbol{A}}_k$.

Remark 1 As can be observed by a straightforward comparison of (2.47) and (2.58), the convergence condition (2.47) is less restrictive than the solution obtained with the approach proposed in Ref. [22], which can be written as

$$
\bar{\alpha}_k \leq \left(\frac{(1 - \zeta)\underline{\sigma} \left(\bar{A}_k P_k \bar{A}_k^T + Q_k \right)}{\bar{\sigma} \left(\bar{A}_k \right)^2 \bar{\sigma} \left(P_k \right)} \right)^{\frac{1}{2}}.
\tag{2.58}
$$

However, (2.47) and (2.58) become equivalent when $E_k \neq \mathbf{0}$, i.e., in all cases when unknown input is considered. This is because of the fact that the matrix $\bar{A}_k$ is singular when $E_k \neq \mathbf{0}$, which implies that $\underline{\sigma} \left(\bar{A}_k \right) = 0$. Indeed, from (2.37),

$$
\bar{A}_k = \bar{G}_k A_k = \left[I - E_k \left[(C_{k+1} E_k)^T C_{k+1} E_k \right]^{-1} \right.
$$

$$
\left. (C_{k+1} E_k)^T C_{k+1} \right] A_k,
\tag{2.59}
$$

and under Assumption 2, it is evident that $\bar{A}_k$ is singular when

$$
E_k \left[(C_{k+1} E_k)^T C_{k+1} E_k \right]^{-1} (C_{k+1} E_k)^T C_{k+1}
$$

is singular. The singularity of the above matrix can be easily shown with the use of (2.27), i.e.,

$$
\text{rank} \left(E_k \left[(C_{k+1} E_k)^T C_{k+1} E_k \right]^{-1} (C_{k+1} E_k)^T C_{k+1} \right)
$$

$$
\leq \min \left[\text{rank}(E_k), \text{rank}(C_{k+1}) \right] = q.
\tag{2.60}
$$

Taking into account the fact that $q < n$, the singularity of $\bar{A}_k$ becomes evident.

Remark 2 It is clear from (2.47) that the bound of $\bar{\alpha}_k$ can be maximised by suitable settings of the instrumental matrices Q_k and R_k. Indeed, Q_k should be selected in such a way as to maximise

$$
\underline{\sigma} \left(\bar{A}_k P_k \bar{A}_k^T + Q_k \right).
\tag{2.61}
$$

To tackle this problem, let us start with a similar solution to the one proposed in Ref. [23], i.e.,

$$
Q_k = \gamma \bar{A}_k P_k \bar{A}_k^T + \delta_1 I,
\tag{2.62}
$$

where $\gamma \geq 0$ and $\delta_1 > 0$. Substituting, (2.62) into (2.61) and taking into account that $\underline{\sigma} \left(\bar{A}_k \right) = 0$, it can be shown that,

$$(1 + \gamma)\underline{\sigma}\left(\bar{A}_k P_k \bar{A}_k^T\right) + \delta_1 I = \delta_1 I. \tag{2.63}$$

Indeed, singularity of $\bar{A}_k$, causes $\underline{\sigma}\left(\bar{A}_k P_k \bar{A}_k^T\right) = 0$, which implies the final result of (2.63). Thus, this solution boils down to the classical approach with constant $Q_k = \delta_1 I$. It is, of course, possible to set $Q_k = \delta_1 I$ with δ_1 large enough. As has been mentioned, the more accurate (near "true" values) the covariance matrices, the better the convergence rate. This means that, in the deterministic case, both matrices should be zero ones. On the other hand, such a solution may lead to the divergence of the observer. To tackle this problem, a compromise between the convergence and the convergence rate should be established. This can be easily done by setting Q_k as

$$Q_k = (\gamma \varepsilon_k^T \varepsilon_k + \delta_1)I, \quad \varepsilon_k = y_k - C_k \hat{x}_k, \tag{2.64}$$

with $\gamma > 0$ and $\delta_1 > 0$ large and small enough, respectively. Since the form of Q_k is established, then it is possible to obtain R_k in such a way as to minimise

$$\bar{\sigma}\left(C_{k+1} P_{k+1/k} C_{k+1}^T + R_{k+1}\right). \tag{2.65}$$

To tackle this problem, let us start with the solution proposed in Refs. [22, 23]:

$$R_{k+1} = \beta_1 C_{k+1} P_{k+1/k} C_{k+1}^T + \delta_2 I, \tag{2.66}$$

with $\beta_1 \geq 0$ and $\delta_2 > 0$. Substituting (2.66) into (2.65) gives

$$(1 + \beta_1)\bar{\sigma}\left(C_{k+1} P_{k+1/k} C_{k+1}^T\right) + \delta_2 I. \tag{2.67}$$

Thus, β_1 in (2.66) should be set so as to minimise (2.67), which implies ($\beta_1 = 0$)

$$R_{k+1} = \delta_2 I, \tag{2.68}$$

with δ_2 small enough.

Second-order Case

From Remark 2 it is clear that the matrix R_k should be set according to (2.68) both in the first- and the second-order case. Indeed, it can be easily observed that its derivation does not depend on the form of Z_k. A significantly different situation takes place in the case of Q_k. Indeed, when (2.64) is employed to set Q_k, then from (2.33) and (2.36) it is evident that P_k is large in the sense of its singular values as well as in the trace. Thus, from (2.38) and (2.42) it is clear that the diagonal entries of β_k should be relatively large. On the other hand, by observing (2.47) it can be concluded that the upper bound of α_k strongly depends on Z_k while

$$\bar{\sigma}\left(\mathbf{Z}_k\right) \leq \bar{\sigma}\left(\bar{\mathbf{A}}_k\right) + \bar{\sigma}\left(\boldsymbol{\beta}_k\right)$$
$$= \bar{\sigma}\left(\bar{\mathbf{A}}_k\right) + \max_{i=1,\ldots,n} |\beta_{i,k}|. \tag{2.69}$$

The above-described situation results in the fact that the upper bound of α_k can be very small (while using (2.64)), which may lead to the divergence of the observer. Thus, the only solution is to set $\mathbf{Q}_k$ in a conventional way, i.e.,

$$\mathbf{Q}_{k+1} = \delta_3 \mathbf{I}, \tag{2.70}$$

with $\delta_3 > 0$ small enough.

2.4 Unscented Kalman Filter

The main objective of this section is to provide a general framework for designing the UIF for non-linear stochastic systems, which is based on the UKF.

As has already been mentioned, state estimation for non-linear stochastic systems is a difficult and important problem for modern fault diagnosis and control systems (see the recent books in the subject area for a complete survey and explanations [2, 3, 24–27]). As can be observed in the literature, the most frequently used approach to state estimation of non-linear stochastic systems is to use the celebrated EKF. However, the linearised non-linear transformations of the state and/or output are reliable only if there is no excessive difference between the local behaviour compared to the original non-linear transformation. If this is not the case, then the EKF will suffer from divergence. However, in the preceding part of this chapter the process and measurement noise matrices are used as instrumental matrices that can significantly improve the convergence performance (see Refs. [2, 18] for a comprehensive survey). Unfortunately, in the stochastic case, $\mathbf{Q}$ and $\mathbf{R}$ have to play their primary role as covariance matrices.

As indicated in Ref. [16], *it is easier to approximate a probability distribution than it is to approximate an arbitrary non-linear function or transformation.*

Bearing in mind this sentence, the idea of an Unscented Transform (UT) was applied along with the celebrated Kalman filter in order to form the UKF. To make the chapter self-contained, the subsequent points will describe the UT and the algorithm of the UKF.

Finally, it should be underlined that the reader is referred to Ref. [16] (and the references therein) for a large number of practical examples showing the superiority of the UKF over the conventional EKF. Thus, the subsequent part of the chapter is focused on developing a new UKF-based scheme rather than showing its superiority over the EKF.

2.4.1 Unscented Transform

The unscented transform boils down to approximating the mean and covariance of the so-called sigma points after the non-linear transformation $h(\cdot)$. The mean and covariance of sigma points are given as $\bar{x}$ and P, while the UT procedure is Ref. [16]

1. Generate k sigma points,

$$X_i, \quad i = 1, \ldots, k, \tag{2.71}$$

   with the mean $\bar{x}$ and covariance P.
2. Obtain a non-linear transformation of each sigma point (cf. Fig. 2.1),

$$X_i^t = h(X_i), \quad i = 1, \ldots, k. \tag{2.72}$$

3. Calculate the weighted mean of the transformed points,

$$\bar{x}^t = \sum_{i=1}^{k} W^i X_i^t. \tag{2.73}$$

4. Calculate the covariance of the transformed points,

$$P^t = \sum_{i=1}^{k} W^i \left[X_i^t - \bar{x}^t \right] \cdot \left[X_i^t - \bar{x}^t \right]^T. \tag{2.74}$$

Note that the sigma points can be generated with various scenarios [16, 17], and one of them will be described in the subsequent point. It should also be mentioned that, in order to provide an unbiased estimate [16], the weights should satisfy

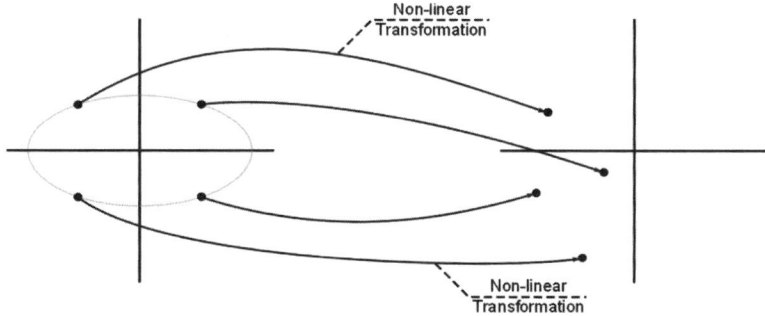

Fig. 2.1 Unscented transform

$$\sum_{i=1}^{k} W^i = 1. \tag{2.75}$$

2.4.2 Principle of the UKF-based UIF

Let us consider a non-linear, discrete-time fault-free system, i.e., (2.1) and (2.2) for $f_k = 0$:

$$x_{k+1} = g(x_k) + h(u_k) + Ed_k + w_k, \tag{2.76}$$

$$y_{k+1} = Cx_{k+1} + v_{k+1}. \tag{2.77}$$

The UKF [17] can be perceived a derivative-free alternative to the extended Kalman filter in the framework of state estimation. The UKF calculates the mean and co-variance of a random variable, which undergoes a non-linear transformation by utilising a deterministic "sampling" approach. Generally, $2n + 1$, *sigma* points are chosen based on a square-root decomposition of the prior covariance. These sigma points are propagated through true nonlinearity, without any approximation, and then a weighted mean and covariance are taken, as described in Sect. 2.4.1.

The presented form of the UKF is based on the general structure of the unknown input observer (2.6) and by taking into account the fact that the output equation (2.77) is linear.

The UKF involves a recursive application of these sigma points to state-space equations. The standard UKF implementation for state estimation uses the following variable definitions:

- $\lambda = 2n(\alpha^2 - 1)$,
- $W_0^m = \frac{\lambda}{n+\lambda}$,
- $W_0^c = \frac{\lambda}{n+\lambda} + 1 - \alpha^2 + \beta$,
- $W_i^m = W_i^c = \frac{1}{2(n+\lambda)}$,
- $\eta = \sqrt{n + \lambda}$,

where W_i^m is a set of scalar weights, and λ and η are scaling parameters. The constant α determines the spread of the sigma points around $\hat{x}$ and is usually set to $10^{-4} \leq \alpha \leq 1$. The constant β is used to incorporate prior knowledge of the distribution (for the Gaussian distribution, $\beta = 2$ is an optimal choice). The UKF algorithm is as follows:

Initialise with

$$\hat{x}_0 = \mathcal{E}[x_0], \quad P_0 = \mathcal{E}[(x_0 - \hat{x}_0)(x_0 - \hat{x}_0)^T], \tag{2.78}$$

For $k \in \{1, \ldots, \infty\}$

Calculate $2n + 1$ sigma points:

$$\hat{X}_{k-1} = [\hat{x}_{k-1} \;\; \hat{x}_{k-1} + \eta S(1), \ldots, \hat{x}_{k-1} + \eta S(n),$$
$$\hat{x}_{k-1} - \eta S(1), \ldots, \hat{x}_{k-1} - \eta S(n)], \tag{2.79}$$

where $S = \sqrt{P_{k-1}}$ and $S(j)$ stands for the jth column of S.
Time update equations:

$$\hat{X}_{i,k|k-1} = \bar{g}\left(\hat{X}_{i,k-1}\right) + \bar{h}(u_k) + \bar{E}y_{k+1}, \quad i = 0, \ldots, 2n, \tag{2.80}$$

$$\hat{x}_{k,k-1} = \sum_{i=0}^{2n} W_i^{(m)} \hat{X}_{i,k|k-1}, \tag{2.81}$$

$$P_{k,k-1} = \sum_{i=0}^{2n} W_i^{(c)} [\hat{X}_{i,k|k-1}$$
$$- \hat{x}_{k,k-1}][\hat{X}_{i,k|k-1} - \hat{x}_{k,k-1}]^T + Q. \tag{2.82}$$

Measurement update equations:

$$P_{y_k y_k} = CP_{k,k-1}C^T + R,$$
$$K_k = P_{k,k-1}C^T P_{y_k y_k}^{-1}, \tag{2.83}$$
$$\hat{y}_{k,k-1} = C\hat{x}_{k,k-1}, \tag{2.84}$$
$$\hat{x}_k = \hat{x}_{k,k-1} + K_k(y_k - \hat{y}_{k,k-1}), \tag{2.85}$$
$$P_k = [I_n - K_k C]P_{k,k-1}. \tag{2.86}$$

2.5 Determination of an Unknown Input Distribution Matrix

As a result of the deliberations presented in the preceding sections, the matrix E should satisfy the following conditions:

$$\text{rank}(CE) = \text{rank}(E) = q, \tag{2.87}$$

where

$$[CE \; CL] \tag{2.88}$$

should be a full rank one, which means that

$$\text{rank}([CE \; CL]) = \min(m, s + q). \tag{2.89}$$

Thus, the set of matrices E satisfying (2.87) and (2.89) is given by

$$\mathbb{E} = \{E \in \mathbb{R}^{n \times q} : \text{rank}(CE) = q \wedge \text{rank}(E) = q \wedge \text{rank}([CE\ CL])$$
$$= \min(m, s + q)\}. \tag{2.90}$$

It should be strongly underlined that $\mathbb{E}$ is not convex, which significantly complicates the problem and limits the spectrum of possible approaches that can be used for settling the determination of the unknown input distribution matrix.

The subsequent part of this section presents a numerical algorithm that can be used for estimating the unknown input distribution matrix E based on a set of input–output measurements $\{(u_k, y_k)\}_{k=1}^{n_t}$.

To settle the problem of numerical estimation of E, the following optimisation criterion is selected:

$$\hat{E} = \arg\min_{E \in \mathbb{E}} J(E) \tag{2.91}$$

with

$$J(E) = \frac{1}{mn_t} \sum_{k=1}^{n_t} z_k^T z_k, \tag{2.92}$$

where z_k stands for the residual defined by (2.7) and $\hat{E}$ is an estimate of E.

It is important to underline that the computation of (2.92) requires the run of the proposed UIF for a given instance of the unknown input distribution matrix E. The computation of the cost function (2.92) is definitely the most time-consuming part of the proposed algorithm. On the other hand, the computation time and the resulting computational burden are not of paramount importance since the proposed algorithm performs off-line. Indeed, only the result of the proposed algorithm, being an estimate of the unknown input distribution matrix E, is utilised on-line for the unknown input decoupling.

The outline of the proposed algorithm is as follows:

Step 1:   Obtain the fault-free input–output data set from the system
$\{(u_k, y_k)\}_{k=1}^{n_t}$.
Step 2:   Initialise the algorithm with some initial value of E satisfying (2.87) and (2.88).
Step 3:   Use an optimisation strategy to find an estimate of E for which (2.92) reaches its minimum and conditions (2.87) and (2.90) are satisfied.

Similarly as in the case of (2.8), i.e., by following with $\tilde{d}_k$ in a similar way as with $\bar{f}_k$ in (2.8), it can be shown that the fault-free residual is

$$z_{k+1} = y_{k+1} - C\hat{x}_{k+1}$$
$$= C\left(\bar{g}(x_k) - \bar{g}(\hat{x}_k) - K(\cdot)\right) + \tilde{d}_k + C\bar{w}_k + v_{k+1}, \tag{2.93}$$

where

$$\tilde{d}_k = C\left[I_n - \hat{E}\left[(C\hat{E})^T C\hat{E}\right]^{-1}(C\hat{E})^T C\right]\bar{d}_k.$$ (2.94)

Alternatively, assuming $\bar{d}_k = Ed_k$, it can be expressed by

$$\tilde{d}_k = C\left[I_n - \hat{E}\left[(C\hat{E})^T C\hat{E}\right]^{-1}(C\hat{E})^T C\right]Ed_k.$$ (2.95)

Following the same line of reasoning as in the proof of Theorem 2.1, it can be shown that for any vector $CEd_k \in \mathrm{col}(C\hat{E})$, the effect of an unknown input $\tilde{d}_k$ will be decoupled from the residual, i.e., $\tilde{d}_k = 0$.

Based on the above deliberations, it seems that an alternative approach is:

Step 0: Obtain the fault-free input–output data set from the system
$\{(u_k, y_k)\}_{k=1}^{n_t}$.
Step 1: Estimate $\bar{d}_k$ for $k = 1, \ldots, n_t$ with, e.g., an augmented UKF.
Step 2: Find a basis of $[\bar{d}_1, \ldots, \bar{d}_{n_t}]$ (e.g. an orthonormal basis), which will constitute an estimate of E.

Apart from the unquestionable appeal of the above algorithm, it does not take into account that the conditions (2.87) and (2.90) must be satisfied. On the other hand, it was empirically proven that, due to the process and measurement noise, accurate estimation of $\bar{d}_k$ (for $k = 1, \ldots, n_t$) is impossible, and hence Step 2 of the above algorithm cannot be realised with expected results.

Thus, the only fruitful conclusion is that an optimal estimate of E is not unique, which will undoubtedly facilitate the performance of the optimisation-based approach presented in the subsequent part of this section.

Taking into account all the above-mentioned difficulties, it is proposed to use the adaptive random search algorithm [2, 28] to solve (2.91). The algorithm has proven to be very reliable in various global optimisation problems, which also justifies its application for this particular task.

The search process of the ARS can be split into two phases. The first phase (variance-selection phase) consists in selecting an element from the sequence

$$\{\sigma^{(i)}\}, \quad i = 1, \ldots, i_{\max},$$ (2.96)

where $\sigma^{(1)}$ stands for an initial standard deviation selected by the designer (forming the covariance matrix $\Sigma = \sigma I_{n \times q}$, where $n \times q$ is the number of elements of E), and

$$\sigma^{(i)} = 10^{(-i+1)}\sigma^{(1)}.$$ (2.97)

In this way, the range of σ ensures both proper exploration properties over the search space and sufficient accuracy of optimum localisation. Larger values of σ decrease the possibility of getting stuck in a local minimum. The second phase (variance-exploration phase) is dedicated to exploring the search space with the use of σ obtained from the first phase and consists in repetitive random perturbation of the best point obtained in the first phase. The scheme of the ARS algorithm is as follows:

0. Input data

- $\sigma^{(1)}$: The initial standard deviation,
- j_{max}: The number of iterations in each phase,
- i_{max}: The number of standard deviations (σ^i) changes,
- k_{max}: The global number of algorithm runs,
- $E^{(0)}$: The initial value of the unknown input distribution matrix.

1. Initialise

(1.1)   Generate $E_{best} \to E^0$, satisfying (2.87) and (2.90), $k \to 1, i \to 1$.

2. Variance-selection phase

(2.1) $j \to 1, E^{(j)} \to E^{(0)}$ and $\sigma^{(i)} \to 10^{(-i+1)}\sigma^{(1)}$.

(2.2)   Perturb $E^{(j)}$ to get a new trial point $E_+^{(j)}$ satisfying (2.87) and (2.90).

(2.3)   If $J(E_+^{(j)}) \leq J(E^{(j)})$ then $E^{(j+1)} \to E_+^{(j)}$
        else $E^{(j+1)} \to E^{(j)}$.

(2.4)   If $J(E_+^{(j)}) \leq J(E_{best})$ then
        $E_{best} \to E_+^{(j)}, i_{best} \to i$.

(2.5)   If $(j \leq j_{max}/i)$ then $j \to j + 1$ and go to (2.2).

(2.6)   If $(i < i_{max})$ then set $i \to i + 1$ and go to (2.1).

3. Variance-exploration phase

(3.1) $j \to 1, E^{(j)} \to E_{best}, i \to i_{best}$
        and $\sigma^{(i)} \to 10^{(-i+1)}\sigma^{(1)}$.

(3.2)   Perturb $E^{(j)}$ to get a new trial point $E_+^{(j)}$ satisfying (2.87) and (2.90).

(3.3)   If $J(E_+^{(j)}) \leq J(E^{(j)})$ then $E^{(j+1)} \to E_+^{(j)}$
        else $E^{(j+1)} \to E^{(j)}$.

(3.4)   If $J(E_+^{(j)}) \leq J(E_{best})$ then $E_{best} \to E_+^{(j)}$.

(3.5)   If $(j \leq j_{max})$ then $j \to j + 1$ and go to Step 3.2.

(3.6)   If $(k \to k_{max})$ then STOP.

(3.7) $k \to k + 1, E^{(0)} \to E_{best}$ and resume from (2.1).

The perturbation phase (the points (2.2) and (3.2) of the algorithm) is realised according to

$$E_+^{(j)} = E^{(j)} + Z, \tag{2.98}$$

where each element of Z is generated according to $\mathcal{N}(0, \sigma^i I)$. When a newly generated $E^{(j)}$ does not satisfy (2.87) and (2.90), then the perturbation phase (2.98) is repeated.

It should also be noted that for some $E^{(j)}$ the proposed UIF may diverge, e.g., due to the loss of observability or a large mismatch with the real system. A simple remedy is to impose a bound (possibly large) ζ on $J(E^{(j)})$, which means that when this bound is exceeded then the UIF is terminated and $J(E^{(j)}) = \zeta$.

2.6 Design of the UIF with Varying Unknown Input Distribution Matrices

The UIF proposed in this section is designed in such a way that it will be able to tackle the problem of automatically changing (or mixing) the influence of unknown input distribution matrices according to system behaviour. In other words, the user can design a number of such matrices in order to cover different operating conditions. Thus, having such a set of matrices, it is possible to design a bank of UIFs and the algorithm should use them to obtain the best unknown input decoupling and state estimation. In order to realise this task, the Interacting Multiple-Model (IMM) approach [29] is used. The subsequent part of this section shows a comprehensive description of the UIF and the IMM.

The IMM solution consists of a filter for each disturbance matrix (corresponding to a particular model of the system), an estimate mixer at the input of the filters, and an estimate combiner at the output of the filters. The IMM works as a recursive estimator. In each recursion it has four steps:

1. Interacting or mixing of model-conditional estimates, in which the input to the filter matched to a certain mode is obtained by mixing the estimates of all filters from the previous time instant under the assumption that this particular mode is in effect at the present time;
2. Model-conditional filtering, performed in parallel for each mode;
3. Model probability update, based on model-conditional innovations and likelihood functions;
4. Estimate combination, which yields the overall state estimate according to the probabilistically weighted sum of updated state estimates of all the filters.

The probability of a mode plays a crucial role in determining the weights in the combination of the state estimates and covariances for the overall state estimate. Figure 2.2 shows the block diagram of the classic IMM algorithm [29], where

- $\hat{x}_{k+1|k+1}$ is the state estimate for time k using measurements through time $(k + 1|k + 1)$ based on N models;
- $\hat{x}^j_{k+1|k+1}$ is the state estimate for time k using measurements through time $(k + 1|k + 1)$ based on model j;
- Λ^j_k is the model likelihood at time k based on model j;
- μ_k is the vector of model probabilities at time k when all the likelihoods Λ^j_k have been considered during model probability update.

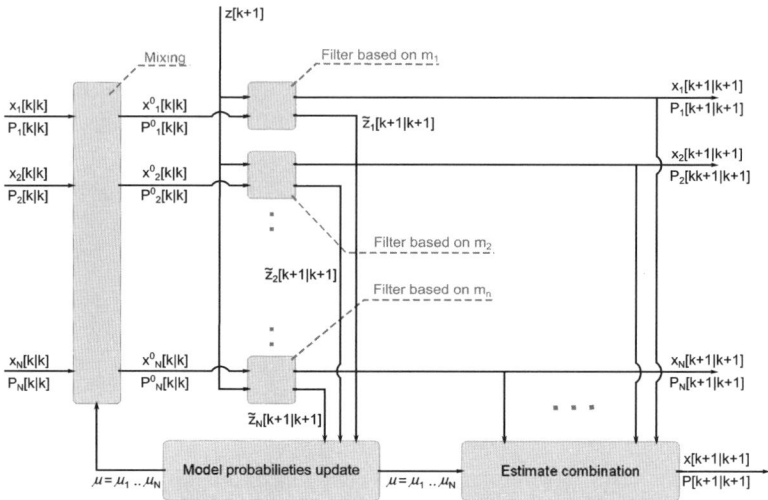

Fig. 2.2 IMM algorithm

With the assumption that model switching is governed by an underlying Markov [29] chain, an interacting mixer at the input of the N filters uses the model probabilities μ_k and the model switching probabilities p_{ij} to compute a mixed (initial or a priori) estimate $\hat{X}_{k|k}^{0j}$ for N filters. The interacting mixer blends the previous state estimates based on N models to obtain new state estimates. The mixing gains $\mu_{k-1|k-1}^{i|j}$ are computed from the preceding model probabilities μ_{k-1}^i and the model switching probabilities p_{ij} in the model probability update.

At the beginning of a filtering cycle, all filters use an a priori mixed estimate $\hat{X}_{k-1|k-1}^{0j}$ and the current measurement y_k to compute a new estimate $\hat{X}_{k|k}^j$ and the likelihood Λ_k^j for the jth model filter. The likelihoods, prior model probabilities, and model switching probabilities are then used by the model probability update to compute the new model probabilities. The overall state estimate $\hat{X}_{k|k}$ is then computed by an estimate combiner with the new state estimates and their probabilities.

The algorithm presented below is a combination of the UIF and the IMM and constitutes a solution to the challenging problem of designing the UIF for a set of predefined unknown input distribution matrices $\{E_j\}_{j=1}^N$.

Step 1:   Mixing state estimates

The filtering process starts with "a priori" state estimates $\hat{X}_{k-1|k-1}^j$, state error covariances $P_{k-1|k-1}$ and the associated probabilities μ_{k-1}^j for each jth filter model corresponding to the jth unknown input distribution matrix. The initial or mixed state estimate and covariance for the jth model at time k is computed as

$$\bar{c}_j = \sum_{i=1}^{N} p_{ij} \mu_{k-1}^{i}, \tag{2.99}$$

$$\mu_{k-1|k-1}^{i|j} = \frac{1}{\bar{c}_j} p_{ij} \mu_{k-1}^{i}, \tag{2.100}$$

$$\hat{X}_{k-1|k-1}^{0j} = \sum_{i=1}^{N} \hat{X}_{k-1|k-1}^{i} \mu_{k-1|k-1}^{i|j}, \tag{2.101}$$

$$P_{k-1|k-1}^{0j} = \sum_{i=1}^{N} [P_{k-1|k-1}^{i} \\ + (\hat{X}_{k-1|k-1}^{i} - \hat{X}_{k-1|k-1}^{0j}) \\ \cdot (\hat{X}_{k-1|k-1}^{i} - \hat{X}_{k-1|k-1}^{0j})^{T}] \mu_{k-1|k-1}^{i|j}. \tag{2.102}$$

p_{ij} is the assumed transition probability for switching from model i to model j, and $\bar{c}_j$ is a normalisation constant. For every state estimate $\hat{X}_{k|k}^{i}$ and $\hat{X}_{k-1|k-1}^{i}$, there is a corresponding covariance $P_{k|k}^{i}$ and $P_{k-1|k-1}^{i}$.

Step 2:   Model-conditioned update

Calculate sigma points (for each jth model):

$$\hat{X}_{k-1}^{j} = [\hat{X}_{k-1|k-1}^{0j} \ \ \hat{X}_{k-1|k-1}^{0j} + \eta\sqrt{P_{k-1|k-1}^{0j}} \\ \hat{X}_{k-1|k-1}^{0j} - \eta\sqrt{P_{k-1|k-1}^{0j}}]. \tag{2.103}$$

Time update (for each jth model):

$$\hat{X}_{i,k|k-1}^{j} = \bar{g}\left(\hat{X}_{i,k-1}^{j}\right) + \bar{h}(u_k) + \bar{E}y_{k+1}, \ i = 0, \ldots, 2n, \tag{2.104}$$

$$\hat{x}_{k,k-1}^{j} = \sum_{i=0}^{2n} W_i^{(m)} \hat{X}_{i,k|k-1}^{j}, \tag{2.105}$$

$$P_{k,k-1}^{j} = \sum_{i=0}^{2n} W_i^{(c)} [\hat{X}_{i,k|k-1}^{j} \\ - \hat{x}_{k,k-1}^{j}][\hat{X}_{i,k|k-1}^{j} - \hat{x}_{k,k-1}^{j}]^{T} + Q. \tag{2.106}$$

Measurement update equations:

$$P_{y_k y_k}^{j} = CP_{k,k-1}^{j}C^{T} + R,$$

$$K_k^{j} = P_{k,k-1}^{j}C^{T}P_{y_k y_k}^{-1 \ (j)}, \tag{2.107}$$

$$\hat{y}_{k,k-1}^{j} = C\hat{x}_{k,k-1}^{j}, \tag{2.108}$$

$$z_k^j = y_k - \hat{y}_{k,k-1}^j, \tag{2.109}$$

$$\hat{x}_{k|k}^j = \hat{x}_{k,k-1}^j + K_k^j z_k^j \tag{2.110}$$

$$P_{k|k}^j = [I_n - K_k C] P_{k|k-1}^j. \tag{2.111}$$

Step 3:   Model likelihood computations

The likelihood of the jth model is computed with the filter residuals z_k^j, the co-variance of the filter residuals $P_{y_k y_k}^j$ and the assumption of the Gaussian statistics. The likelihood of the jth model and model probabilities update are as follows:

$$\Lambda_k^j = \frac{1}{\sqrt{|2\pi P_{y_k y_k}^j|}} \exp\left[-0.5(z_k^j)^T (P_{y_k y_k}^j)^{-1} z_k^j\right],$$

$$c = \sum_{i=1}^N \Lambda_k^i \bar{c}_i,$$

$$\mu_k^j = \frac{1}{c} \Lambda_k^j \bar{c}_j.$$

Step 4:   Combination of state estimates

The state estimate $\hat{x}_{k|k}$ and the covariance $P_{k|k}$ for the IMM filter are obtained from a probabilistic sum of the individual filter outputs,

$$\hat{x}_{k|k} = \sum_{i=1}^N \hat{x}_{k|k}^i \mu_k^i,$$

$$P_{k|k} = \sum_{i=1}^N \mu_k^i \left[P_{k|k}^i + (\hat{x}_{k|k}^i - \hat{x}_{k|k})(\hat{x}_{k|k}^i - \hat{x}_{k|k})^T\right].$$

2.7 Illustrative Examples

The objective of the subsequent part of this section is to examine the proposed approaches with two sample systems, i.e., an induction motor and a two-tank system. In particular, the way of determining unknown input distribution matrix and the "switching" of these matrices will be illustrated with an induction motor. The two-tank system will be employed to show the performance of the proposed approach with respect to fault detection and isolation. The final part of this section shows a comparison between the first- and the second-order EUIO. An empirical comparison is realised with the model of an induction motor.

2.7.1 Estimation of E for an Induction Motor

The purpose of this section is to show the reliability and effectiveness of the proposed EUIO. The numerical example considered here is a fifth-order two-phase non-linear model of an induction motor, which has already been the subject of a large number of various control design applications (see Ref. [22] and the references therein).The complete discrete-time model in a stator-fixed (a, b) reference frame is

$$x_{1,k+1} = x_{1,k} + h\left(-\gamma x_{1k} + \frac{K}{T_r}x_{3k} + Kpx_{5k}x_{4k} + \frac{1}{\sigma L_s}u_{1k}\right), \qquad (2.112)$$

$$x_{2,k+1} = x_{2,k} + h\left(-\gamma x_{2k} - Kpx_{5k}x_{3k} + \frac{K}{T_r}x_{4k} + \frac{1}{\sigma L_s}u_{2k}\right), \qquad (2.113)$$

$$x_{3,k+1} = x_{3,k} + h\left(\frac{M}{T_r}x_{1k} - \frac{1}{T_r}x_{3k} - px_{5k}x_{4k}\right), \qquad (2.114)$$

$$x_{4,k+1} = x_{4,k} + h\left(\frac{M}{T_r}x_{2k} + px_{5k}x_{3k} - \frac{1}{T_r}x_{4k}\right), \qquad (2.115)$$

$$x_{5,k+1} = x_{5,k} + h\left(\frac{pM}{JL_r}(x_{3k}x_{2k} - x_{4k}x_{1k}) - \frac{T_L}{J}\right), \qquad (2.116)$$

$$y_{1,k+1} = x_{1,k+1}, \quad y_{2,k+1} = x_{2,k+1}, \qquad (2.117)$$

where $x_k = [x_{1,k}, \ldots, x_{n,k}]^T = [i_{sak}, i_{sbk}, \psi_{rak}, \psi_{rbk}, \omega_k]^T$ represents the currents, the rotor fluxes, and the angular speed, respectively, while $u_k = [u_{sak}, u_{sbk}]^T$ is the stator voltage control vector, p is the number of the pairs of poles, and T_L is the load torque. The rotor time constant T_r and the remaining parameters are defined as

$$T_r = \frac{L_r}{R_r}, \quad \sigma = 1 - \frac{M^2}{L_sL_r}, \quad K = \frac{M}{\sigma L_sL_r}, \quad \gamma = \frac{R_s}{\sigma L_s} + \frac{R_rM^2}{\sigma L_sL_r^2}, \qquad (2.118)$$

where R_s, R_r and L_s, L_r are stator and rotor per-phase resistances and inductances, respectively, and J is the rotor moment inertia.

The numerical values of the above parameters are as follows: $R_s = 0.18 \, \Omega$, $R_r = 0.15 \, \Omega$, $M = 0.068 \, H$, $L_s = 0.0699 \, H$, $L_r = 0.0699 \, H$, $J = 0.0586 \, kgm^2$, $T_L = 10 \, Nm$, $p = 1$, and $h = 0.1 \, ms$. The input signals are

$$u_{1,k} = 350\cos(0.03k), \quad u_{2,k} = 300\sin(0.03k). \qquad (2.119)$$

Let us assume that the unknown input and its distribution matrix have the following form:

$$E = [1.2, 0.2, 2.4, 1, 1.6]^T, \qquad (2.120)$$

$$d_k = 0.3\sin(0.5\pi k)\cos(0.03\pi k), \qquad (2.121)$$

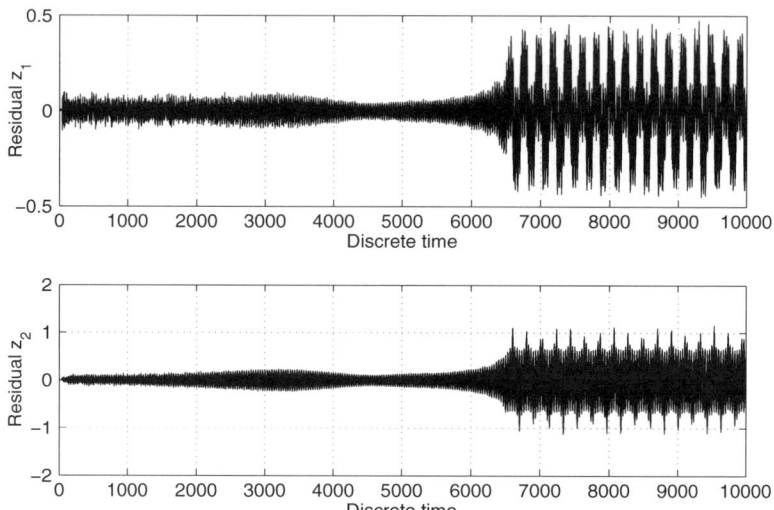

Fig. 2.3 Residuals for a randomly selected E

while the noise covariance matrices are $Q = 10^{-5}I$ and $R = 10^{-5}I$. Note that the small values of the process and measurement noise are selected in order to clearly portray the effect of an unknown input. Figure 2.3 shows the residual z_k for randomly selected E. From these results, it is evident that the estimation quality is very low and hence the residual is significantly different from zero, which may lead to a decrease in the fault detection abilities.

In order to prevent such a situation, the algorithm presented in Sect. 2.5 was utilised with the following settings:

- $\sigma^{(1)}$: The initial standard deviation,
- $j_{max} = 20$: The number of iterations in each phase,
- $i_{max} = 5$: The number of standard deviations (σ^i) changes,
- $k_{max} = 50$,
- $E^{(0)}$: Randomly selected.

The performance of the algorithm was tested for a set of $\sigma^{(1)}$, i.e., $\{1, 2, 3, 4, 5\}$. Note that $k_{max} = 50$, which means that each run of the algorithm was performed 50 times. As a result, the mean and the standard deviation of the resulting $J(E)$ (cf. (2.92)) for each setting of $\sigma^{(1)}$ was calculated. The mean of $J(E)$ is presented in Fig. 2.4, while its standard deviation is portrayed in Fig. 2.5.

From these results, it is evident that the smallest mean and standard deviation are obtained for $\sigma^{(1)} = 3$. This, of course, does not mean that this is a particular value $\sigma^{(1)} = 3$, which should be the best one for each example. However, it can be easily observed that, for other $\sigma^{(1)}$, i.e., $\{1, 2, 4, 5\}$, the mean and standard deviation are

Fig. 2.4 Mean of $J(E)$ for $\sigma^{(1)} = 1, \ldots, 5$

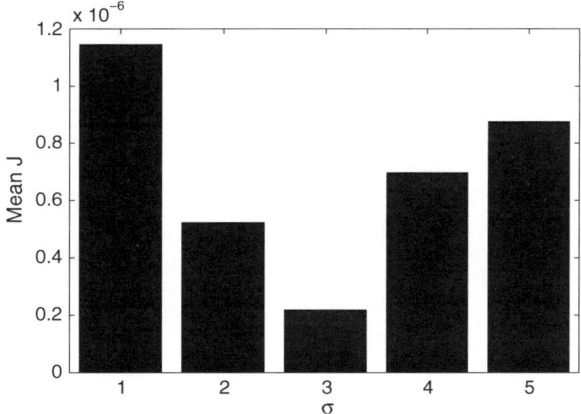

Fig. 2.5 Standard deviation of $J(E)$ for $\sigma^{(1)} = 1, \ldots, 5$

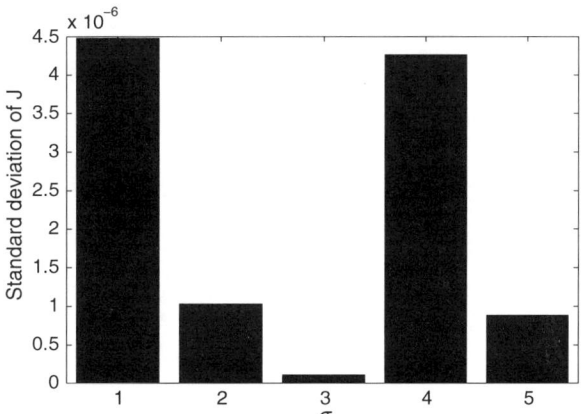

also very small. Numerous numerical experiments confirm this property, i.e., this means that the proposed algorithm is not extremely sensitive to the initial value of $\sigma^{(1)}$.

As mentioned in the preceding part of the chapter, the matrix E, which is able to decouple the unknown input, is not unique. Indeed, the estimate of E, for which $J(E)$ reaches its minimum, is

$$\hat{E} = [0.3651, 0.0609, 0.7303, 0.3043, 0.4869]^{T}. \tag{2.122}$$

Figure 2.6 presents the residual for the obtained estimate. A direct comparison of Figs. 2.3 and 2.6 clearly shows the profits that can be gained while using the proposed algorithm.

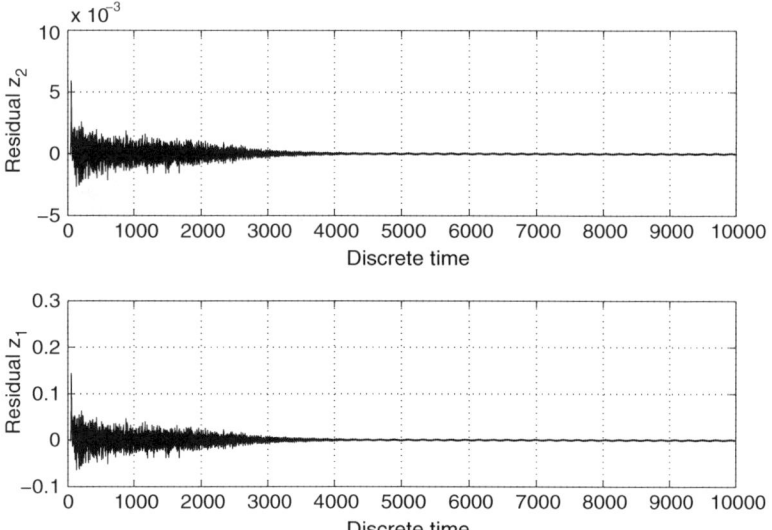

Fig. 2.6 Residuals for the estimated E

2.7.2 Varying E Case

Let us reconsider an example presented in the previous section. The unknown input is defined, as previously, by (2.121), but three different settings of the unknown input distribution matrix E^j were employed during system simulation (the simulation time was 10,000 samples):

$$E^1 = [1.2, 0.2, 2.4, 1, 1.6]^T \text{ for } 0 \leq k < 2,500,$$
$$E^2 = [0.2, 1.2, 2.4, 1, 1.6]^T \text{ for } 2,500 \leq k < 5,000$$
$$\text{and } 7,500 \leq k < 10,000,$$
$$E^3 = [2.1, 2.1, 2.1, 2.1, 2.1]^T \text{ for } 5,000 \leq k < 7,500.$$

Contrary to the above-described simulation scenario, it was assumed that the set of unknown input distribution matrices for the UIF is composed of

$$E^1 = [0.2, 1.2, 2.4, 1, 1.6]^T,$$
$$E^2 = [0, 0.2, 2.4, 1, 0]^T,$$
$$E^3 = [2.1, 2.1, 2.1, 2.1, 2.1]^T,$$
$$E^4 = [1, 2, 3, 1, 0]^T,$$
$$E^5 = [1.2, 0.2, 2.4, 1, 1.6]^T.$$

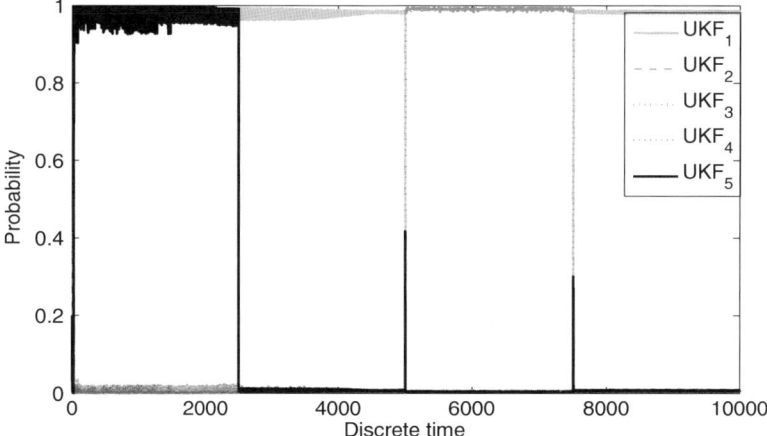

Fig. 2.7 Model probabilities

This means that E^2 and E^4 should not be used by the UIF, while E^1, E^3 and E^5 should be appropriately switched.

Figure 2.7 shows model probabilities corresponding to the five unknown input distribution matrices. From these results, it is evident that the instrumental matrices E^1, E^3 and E^5 were switched correctly. Moreover, probabilities corresponding to E^2 and E^4 are very low.

2.7.3 Fault Detection and Isolation of a Two-Tank System

The main objective of this section is to show that the UIF can also be effectively applied for fault detection and isolation. In this case, the unknown input is suitably used for designing the required fault isolation performance. The system considered consists of two cylindrical tanks of the same diameter. They are linked to each other through a connecting cylindrical pipe (Fig. 2.8). The two-tank system can be perceived as a Single-Input Multi-Output (SIMO) system, where the input u is the water flow through the pump, while the outputs y_1 and y_2 are water levels in the first and the second tank, respectively.

It is assumed that the system considered can be affected by the following set of faults:

actuator fault: f_1 pump lost-of-effectiveness or leakage from the pump pipe,
process faults: f_2 clogged connecting cylindrical pipe,
sensor faults: f_3 water level sensor fault of the first tank, f_4 water level sensor fault of the second tank.

Fig. 2.8 Schematic diagram
of a two-tank system

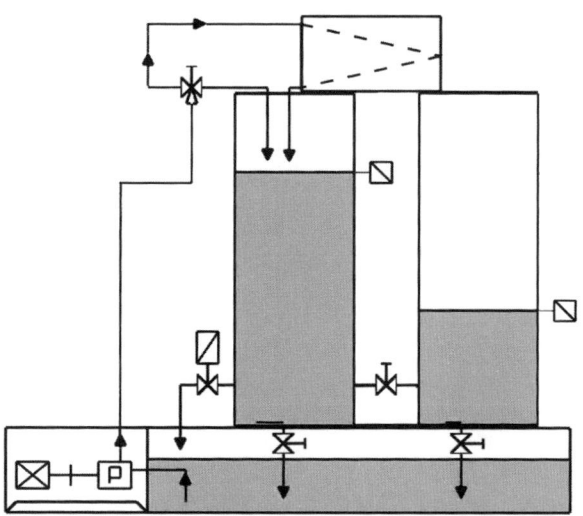

Once the fault description is provided, then a complete system description can be
given as follows:

$$x_{k+1} = g(x_k) + h(u_k) + L_1 f_{a,k} + w_k, \qquad (2.123)$$

$$y_{k+1} = C x_{k+1} + L_2 f_{s,k+1} + v_{k+1}, \qquad (2.124)$$

where

$$g(x_k) = \begin{bmatrix} -h\frac{K_1}{A_1}\sqrt{x_{1,k} - x_{2,k}} + x_{1,k} \\ h\frac{K_1}{A_2}\sqrt{x_{1,k} - x_{2,k}} - h\frac{K_2}{A_2}\sqrt{x_{2,k}} + x_{2,k} \end{bmatrix}, \qquad (2.125)$$

$$h(u_k) = \left[h\frac{1}{A_1}u_k, 0 \right]^T, \qquad (2.126)$$

$$L_1 = \begin{bmatrix} -\frac{h}{A_1} & \frac{h}{A_1} \\ 0 & \frac{-h}{A_2} \end{bmatrix}, \qquad (2.127)$$

$$L_2 = \begin{bmatrix} 1 & 0 \\ 0 & 1 \end{bmatrix}, \quad C = I, \qquad (2.128)$$

$$f_{a,k} = \left[f_{1,k}, \sqrt{x_{1,k} - x_{2,k}} f_{2,k} \right]^T, \quad f_{s,k} = \left[f_{3,k}, f_{4,k} \right]^T,$$

where $x_{1,k}$ and $x_{2,k}$ are water levels in the first and the second tank, respectively,
A_1, A_2 stand for the cross-sections of the tanks, K_1 denotes the cross-section of the

connecting pipe, K_2 is the cross-section of the outflow pipe from the second tank, and h is the sampling time.

The objective of the subsequent part of this section is to design UIF-based diagnostic filter which will make it possible to detect and isolate the above mentioned faults.

Filter 1: In order to make the residual insensitive to f_1, it is proposed to use the developed UIF with the following settings:

$$E = L_1^1, \quad d_k = f_{1,k}, \quad L = L_1^2, \quad f_k = f_{2,k}, \quad C_k = [1,\ 0], \qquad (2.129)$$

where L_1^i stands for the ith column of L_1. It is straightforward to examine that the conditions (2.87) and (2.90) are satisfied, which means that the observer while remaining insensitive to $f_{1,k}$ while it will remain sensitive to $f_{2,k}$.

Filter 2: Similarly as in the *Filter 1* case, the residual generated by the *Filter 2* should be insensitive to $f_{2,k}$,

$$E = L_1^2, \quad d_k = f_{2,k}, \quad L = L_1^1, \quad f_k = f_{1,k}, \quad C_k = [1,\ 0]. \qquad (2.130)$$

It is straightforward to examine that conditions (2.87) and (2.90) are satisfied, which means that the observer will be insensitive to $f_{2,k}$ while while remaining sensitive to $f_{1,k}$.

Filter 3: The filter should be insensitive to $f_{3,k}$ while sensitive to $f_{4,k}$. This can be realised using the conventional UKF with

$$C = [0,\ 1]. \qquad (2.131)$$

Filter 4: The filter should be insensitive to $f_{4,k}$ while it should be sensitive to $f_{3,k}$. This can be realised using the conventional UKF with

$$C = [1,\ 0], \qquad (2.132)$$

The main objective of this section is to show the testing results obtained with the proposed approach. To tackle this problem, a Matlab-based simulator of a two-tank system was implemented. The simulator is able to generate the data for normal as well as for all faulty conditions $(f_1, \ldots, f_4)$ being considered. The filter-based fault diagnosis scheme was also implemented using Matlab. As a result, a complete scheme that is able to validate the performance of the proposed fault diagnosis strategy was developed. It should be also pointed out that the simulations were carried out using the following numerical parameters: $u_k = 2.56$, $h = 0.1$, $A_1 = 4.2929$, $A_2 = 4.2929$, $K_1 = 0.3646$, $K_2 = 0.2524$.

All fault scenarios where generated according to the following rule:

$$f_{i,k} = \begin{cases} \neq 0 & k = 300, \ldots, 400 \\ 0 & \text{otherwise} \end{cases} \quad i = 1, \ldots, 4.$$

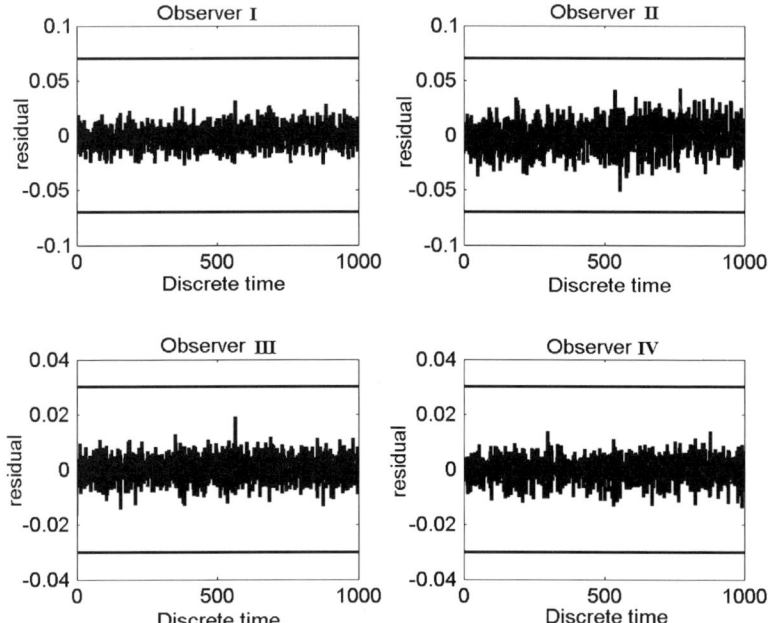

Fig. 2.9 Residuals for the fault-free case

Moreover, y_1 and y_2 were corrupted by measurement noise generated according to the normal distribution, i.e., $\mathcal{N}(\mathbf{0}, \mathrm{diag}(0.01, 0.01))$. Thus, the following settings of the instrumental matrices were employed: $\boldsymbol{R} = 0.1\boldsymbol{I}$ and $\boldsymbol{Q} = 0.1\boldsymbol{I}$.

Figure 2.9 portrays the residual obtained with the four filters for the fault-free case. As can be observed, all of them are very close to zero.

Figures 2.10, 2.11, 2.12 and 2.13 present the residuals for the faults f_1 to f_4 obtained with the four filters.

The results are summarised in the form of a diagnostic table presented as Table 2.1.

It should be noticed that the residuals generated by *Filter 3* and *Filter 4* are insensitive to f_1 and f_2. Such a situation is caused by the fact that observers use feedback from the system output and hence some damping effects may arise. This is the case in the presented situation. On the other hand, it was observed that the results of experiments can be consistent with the theoretical expectations when there is no measurement noise, but this is a rather unreal situation. Irrespective of the presented results, the faults can still be isolated because they have unique signatures.

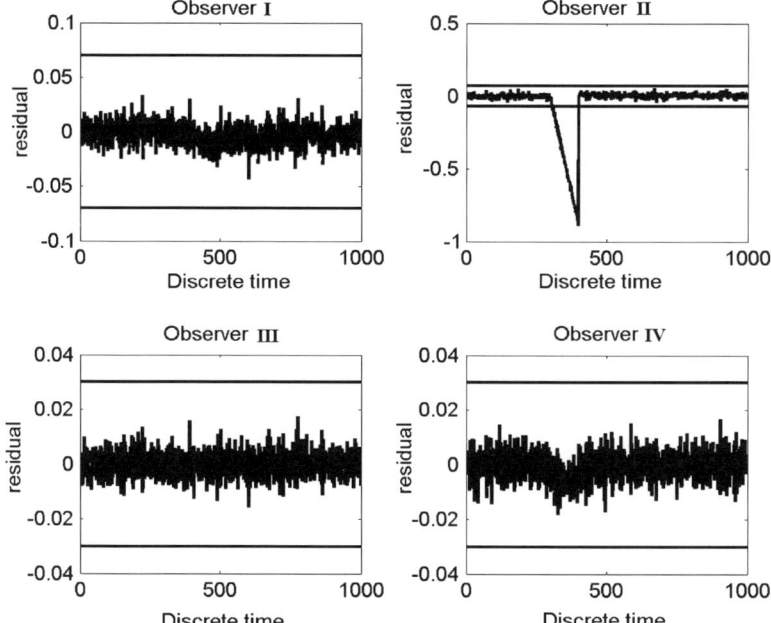

Fig. 2.10 Residuals for the fault f_1

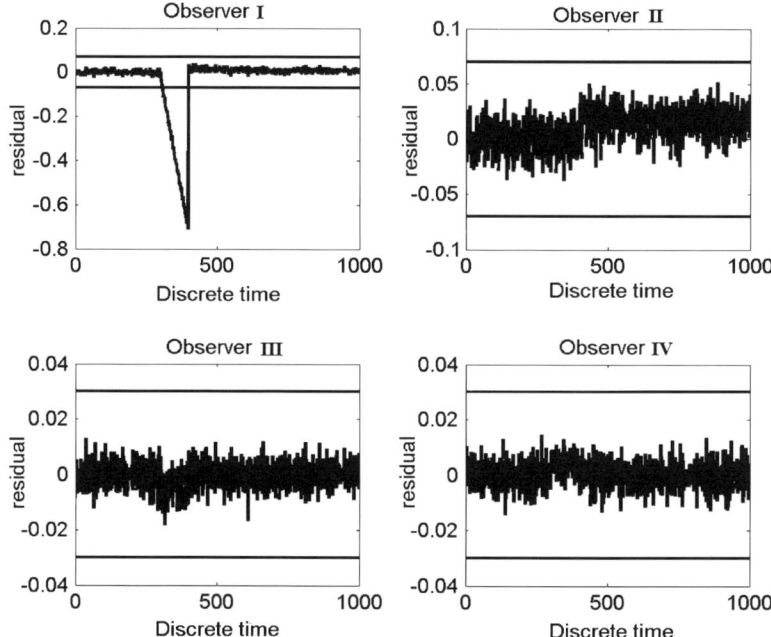

Fig. 2.11 Residuals for the fault f_2

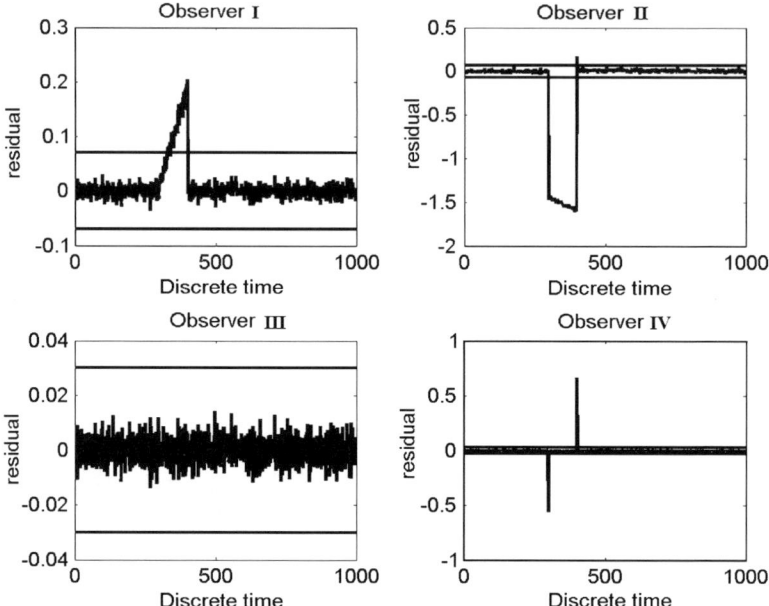

Fig. 2.12 Residuals for the fault f_3

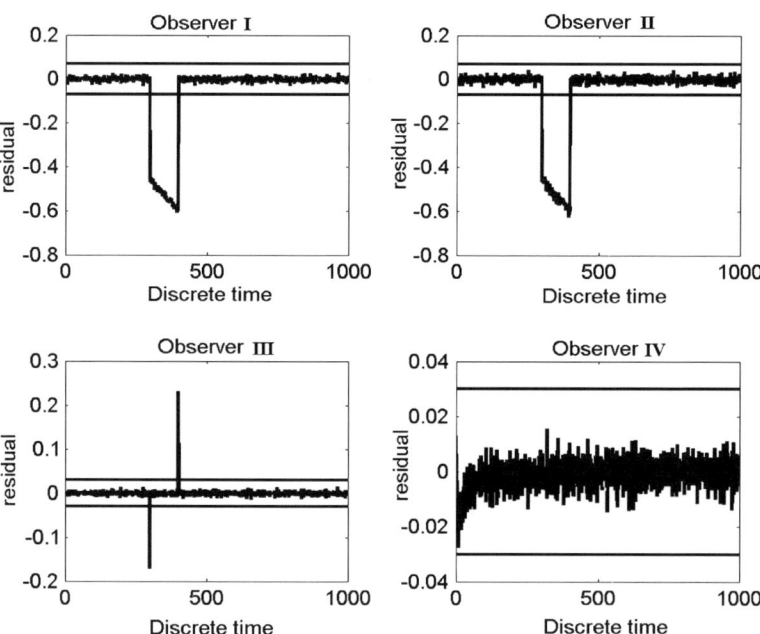

Fig. 2.13 Residuals for the fault f_4

Table 2.1 Diagnostic table

Filter	f_1	f_2	f_3	f_4
Filter 1	0	1	1	1
Filter 2	1	0	1	1
Filter 3	0	0	0	1
Filter 4	0	0	1	0

2.7.4 First- Versus Second-order EUIO

The main objective of this section is to perform a comprehensive study regarding the first- and the second-order EUIO.

Let us reconsider the induction motor described by (2.112)–(2.117). Let $\mathbb{X}$ be a bounded set denoting the space of the possible variations of the initial condition x_0:

$$\mathbb{X} = \{[-276, 279] \times [-243, 369] \times S_{15}^{(2)}(\mathbf{0})$$
$$\times [-11, 56]\} \subset \mathbb{R}^5, \tag{2.133}$$

where $S_r^{(n)}(c) = \{x \in \mathbb{R}^n : \|x - c\|_2 \leq r\}$, $r = 15$. Let us assume that each initial condition of the system x_0 is equally probable, i.e.

$$pr(x_0) = \begin{cases} \frac{1}{m(\mathbb{X})} & \text{for } x_0 \in \mathbb{X}, \\ 0 & \text{otherwise,} \end{cases}$$

where $m(\mathbb{A})$ is the Lebesgue measure of the set $\mathbb{A}$. Moreover, the following three observer configurations were considered:

Case 1:   First-order EUIO with:

$$R_k = 0.1I_2,$$
$$Q_k = 0.1I_5.$$

Case 2:   First-order EUIO with:

$$R_k = 0.1I_2,$$
$$Q_k = 10^3 \varepsilon_k^T \varepsilon_k I_5 + 0.001I_5.$$

Case 3:   Second-order EUIO with:

$$R_k = 0.1I_2,$$
$$Q_k = 0.1I_5.$$

In order to validate the performance of the observers, each of them was run for $N = 1000$ randomly selected initial conditions $x_0 \in \mathbb{X}$ and then the following quality index was calculated:

Fig. 2.14 Average norm of
the state estimation error

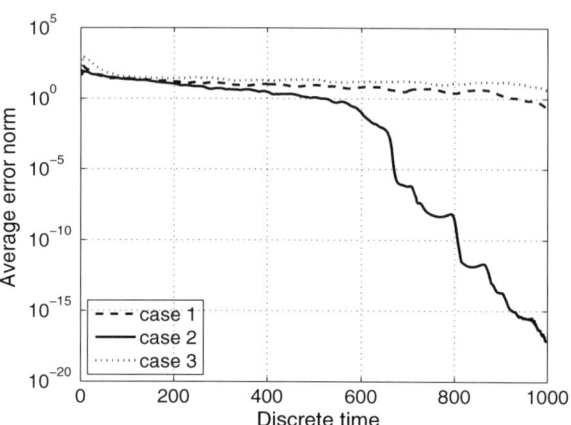

$$Q = \frac{1}{N} \sum_{i=1}^{N} \frac{\|e_{n_t}^i\|_2}{\|e_0^i\|_2}, \qquad (2.134)$$

where $n_t = 1000$. This quality index describes an average ration between the norm
of the initial state estimation error e_0 and the norm of the state estimation estimation
error achieved after n_t iterations. The following results were obtained:

Case 1: $Q = 0.0023$,
Case 2: $Q = 9.6935 \cdot 10^{-17}$,
Case 3: $Q = 0.0385$.

It is clear that the observer of Case 2 provides the best results. This can also be
observed in Fig. 2.14, which presents the evolution of an average norm of the state
estimation error. The main reason why the second-order EUIO does not provide good
results is that it is very sensitive to the initial state estimation error and to the initial
value of P_k. This follows from the fact that s_k is calculated using the approximation
(2.38) instead of an exact form:

$$s_{i,k} = \frac{1}{2} \left[e_k^T \left. \frac{\partial \bar{g}_i (x_k)^2}{\partial x_k^2} \right|_{x_k = \hat{x}_k} e_k \right], \quad i = 1, \ldots, n.$$

2.8 Concluding Remarks

The main objective of this chapter was to present three different approaches that can
be used for designing unknown input observers and filters for non-linear discrete-
time systems. In particular, a system description was provided, which covers a large
class of non-linear systems, and a general rule for decoupling the unknown input

was portrayed. The chapter provides a rule for checking if the observer/filter will not decouple the effect of a fault. This unappealing phenomenon may lead to undetected faults, which may have a serious impact on the performance of the system being controlled and diagnosed. Subsequently, an approach for designing the so-called first- and second-order extended unknown input observer was provided. The proposed approach is based on a general extended Kalman filter framework and can be applied for non-linear deterministic systems. It was shown, with the help of the Lyapunov method, that such a linearisation-based technique is convergent under certain conditions. To tackle this task, a novel structure and design procedure of the EUIO were proposed. Another approach was proposed for non-linear stochastic framework, bearing in mind that Ref. [16] *it is easier to approximate a probability distribution than it is to approximate an arbitrary non-linear function or transformation.* The unscented Kalman filter formed a base for the development of an unknown input filter. Based on the UIF, an algorithm for estimating unknown input distribution matrix was proposed. Another important contribution of this chapter was the development of the UIF that is able to switch the unknown input distribution matrices according to the working conditions. This task was realised with the interacting multiple model algorithm. The final part of the chapter presented comprehensive case studies regarding practical application of the proposed approaches. These examples are an induction motor and a two-tank systems. In particular, based on the example with the induction motor, the strategies for determining the unknown input distribution matrix and the case with a set of predefined unknown input distribution matrices were examined. The same example (within a deterministic framework) was utilised to perform a comparison between the first- and second-order extended unknown input observer. The abilities regarding fault detection and isolation were illustrated with the two-tank system. In all the cases, the proposed approaches exhibit their practical usefulness.

References

1. J. Korbicz, J. Kościelny, Z. Kowalczuk, W. Cholewa (eds.), *Fault Diagnosis. Models, Artificial Intelligence, Applications* (Springer-Verlag, Berlin, 2004)
2. M. Witczak, *Modelling and Estimation Strategies for Fault Diagnosis of Non-linear Systems* (Springer-Verlag, Berlin, 2007)
3. S.X. Ding, *Model-Based Fault Diagnosis Techniques: Design Schemes. Algorithms and Tools* (Springer-Verlag, Berlin, 2008)
4. V. Puig, Fault diagnosis and fault tolerant control using set-membership approaches: application to real case studies. Int. J. Appl. Math. Comput. Sci. **20**(4), 619–635 (2010)
5. S. Tong, G. Yang, W. Zhang, Observer-based fault-tolerant control againts sensor failures for fuzzy systems with time delays. Int. J. Appl. Math. Comput. Sci. **21**(4), 617–628 (2011)
6. K. Kemir, F. Ben Hmida, J. Ragot, M. Gossa, Novel optimal recursive filter for state and fault estimation of linear systems with unknown disturbances. Int. J. Appl. Math. Comput. Sci. **21**(4), 629–638 (2011)
7. P.M. Frank, T. Marcu, Diagnosis strategies and systems. principles, fuzzy and neural approaches, in *Intelligent Systems and Interfaces*, ed. by H.N. Teodorescu, D. Mlynek, A. Kandel, H.J. Zimmermann (Kluwer Academic Publishers, Boston, 2000)

8. S.H. Wang, E.J. Davison, P. Dorato, Observing the states of systems with unmeasurable disturbances. IEEE Trans. Autom. Control **20**(5), 716–717 (1975)
9. H. Hammouri, P. Kabore, S. Othman, J. Biston, Failure diagnosis and nonlinear observer. Application to a hydraulic process. J. Franklin Inst. **339**(4–5), 455–478 (2002)
10. H. Hammouri, P. Kabore, S. Othman, J. Biston, Observer-based approach to fault detection and isolation for nonlinear systems. IEEE Trans. Autom. Control **44**(10), 1879–1884 (1999)
11. R. Kabore, H. Wang, Design of fault diagnosis filters and fault tolerant control for a class of nonlinear systems. IEEE Trans. Autom. Control **46**(11), 1805–1809 (2001)
12. W. Chen, A.Q. Khan, M. Abid, S.X. Ding, Integrated design of observer-based fault detection for a class of uncertain non-linear systems. Int. J. Appl. Math. Comput. Sci. **21**(4), 619–636 (2011)
13. D. Koenig, S. Mammar, Design of a class of reduced unknown inputs non-linear observer for fault diagnosis, in *Proceedings of American Control Conference*, ACC, Arlington, USA, 2002
14. A.M. Pertew, H.J. Marquez, Q. Zhao, $\mathcal{H}_\infty$ synthesis of unknown input observers for non-linear lipschitz systems. Int. J. Control **78**(15), 1155–1165 (2005)
15. J. Chen, R.J. Patton, *Robust Model Based Fault Diagnosis for Dynamic Systems* (Kluwer Academic Publishers, London, 1999)
16. S.J. Julier, J.K. Uhlmann, Unscented filtering and estimation. Proc. IEEE **92**(3), 401–422 (2004)
17. R. Kandepu, B. Foss, L. Imsland, Applying the unscented Kalman filter for nonlinear state estimation. J. Process Control **18**(7–8), 753–768 (2008)
18. M. Witczak, P. Pretki, Design of an extended unknown input observer with stochastic robustness techniques and evolutionary algorithms. Int. J. Control **80**(5), 749–762 (2007)
19. F.J. Uppal, R.J. Patton, M. Witczak, A neuro-fuzzy multiple-model observer approach to robust fault diagnosis based on the DAMADICS benchmark problem. Control Eng. Pract. **14**(6), 699–717 (2006)
20. M. Witczak, J. Korbicz, R. Jozefowicz, Design of unknown input observers for non-linear stochastic systems and their application to robust fault diagnosis. Control Cybern. **42**(1), 227–256 (2013)
21. M. Witczak, P. Pretki, J. Korbicz, Design and convergence analysis of first- and second-order extended unknown input observers, in *Methods and Models in Automation and Robotics—MMAR 2007: Proceedings of the 13th IEEE/IFAC International Conference*, Szczecin, Poland, 2007, pp. 833–838 (CD-ROM)
22. M. Boutayeb, D. Aubry, A strong tracking extended Kalman observer for nonlinear discrete-time systems. IEEE Trans. Autom. Control **44**(8), 1550–1556 (1999)
23. L.Z. Guo, Q.M. Zhu, A fast convergent extended Kalman observer for non-linear discrete-time systems. Int. J. Syst. Sci. **33**(13), 1051–1058 (2002)
24. G. Ducard, *Fault-tolerant Flight Control and Guidance Systems: Practical Methods for Small Unmanned Aerial Vehicles* (Springer-Verlag, Berlin, 2009)
25. R. Iserman, *Fault Diagnosis Applications: Model Based Condition Monitoring, Actuators, Drives, Machinery, Plants, Sensors, and Fault-Tolerant Systems* (Springer-Verlag, Berlin, 2011)
26. M. Mahmoud, J. Jiang, Y. Zhang, *Active Fault Tolerant Control Systems: Stochastic Analysis and Synthesis* (Springer-Verlag, Berlin, 2003)
27. H. Noura, D. Theilliol, J. Ponsart, A. Chamseddine, *Fault-Tolerant Control Systems: Design and Practical Applications* (Springer-Verlag, Berlin, 2003)
28. E. Walter, L. Pronzato, *Identification of Parametric Models from Experimental Data* (Springer, London, 1996)
29. H.A.P. Blom, Y. Bar-Shalom, The interacting multiple model algorithm for systems with markovian switching coefficients. IEEE Trans. Autom. Control **33**(8), 780–783 (1988)

Chapter 3
Neural Network-based Approaches to Fault Diagnosis

Apart from the unquestionable effectiveness of the approaches presented in the preceding chapter, there are examples for which fault directions are very similar to that of an unknown input. This may lead to a situation in which the effect of some faults is minimised and hence they may be impossible to detect. Other approaches that make use of the idea of an unknown input also inherit these drawbacks, e.g., robust parity relations approaches [1]. An obvious way to tackle such a challenging problem is to use different description of model uncertainty. It is important to note that the parameters of the model underlying such a fault detection scheme do not necessarily have to have physical meaning. This considerably extends the spectrum of candidate models that can be used for design purposes. The main objective of this chapter is to show how to use artificial neural networks in such a robust fault detection scheme. Contrary to the industrial applications of neural networks that are presented in the literature (see, e.g, [2] and the references therein), the task of designing a neural network is defined in this chapter in such a way as to obtain a model with a possibly small uncertainty. Indeed, the approaches presented in the literature try to obtain a model that is best suited to a particular data set. This may result in a model with a relatively large uncertainty. Degraded performance of fault diagnosis constitutes a direct consequence of using such models.

Taking into account the above discussion, the chapter is organised as follows. Section 3.1 extends the general ideas of experimental design to neural networks. In particular, one objective of this section is to show how to describe model uncertainty of a neural network using the statistical framework. Another objective is to propose algorithms that can be used for developing an optimal experimental design which makes it possible to obtain a neural network with a possibly small uncertainty. The final objective is to show how to use the obtained knowledge about model uncertainty for robust fault diagnosis. The approach presented in this section is based on a static neural network, i.e., the multi-layer perceptron. It should also be pointed out that the results described in this section are based on [3–5].

An approach that can utilise either a static or a dynamic model structure is described in Sect. 3.2. This strategy is based on a similar idea as that of Sect. 3.1,

M. Witczak, *Fault Diagnosis and Fault-Tolerant Control Strategies for Non-Linear Systems*,   57
Lecture Notes in Electrical Engineering 266, DOI: 10.1007/978-3-319-03014-2_3,
© Springer International Publishing Switzerland 2014

but instead of using a statistical description of model uncertainty a deterministic bounded-error [6, 7] approach is employed. In particular, one objective is to show how to describe model uncertainty of the so-called Group Method of Data Handling (GMDH) neural network. Another objective is to show how to use the obtained knowledge about model uncertainty for robust fault diagnosis. Finally, it should be pointed out that the presented results are based on [5, 8, 9].

3.1 Robust Fault Detection with the Multi-Layer Perceptron

Let us consider a feed-forward neural network given by the following equation:

$$y_{M,k} = P^{(l)} g\left(P^{(n)} u_k \right), \tag{3.1}$$

while $g(\cdot) = [g_1(\cdot), \ldots, g_{n_h}(\cdot), 1]^T$, while $g_i(\cdot) = g(\cdot)$ is a non-linear differentiable activation function,

$$P^{(l)} = \begin{bmatrix} p^{(l)}(1)^T \\ \vdots \\ p^{(l)}(m)^T \end{bmatrix}, \quad P^{(n)} = \begin{bmatrix} p^{(n)}(1)^T \\ \vdots \\ p^{(n)}(n_h)^T \end{bmatrix}, \tag{3.2}$$

are matrices representing the parameters (weights) of the model, and n_h is the number of neurons in the hidden layer. Moreover, $u_k \in \mathbb{R}^{r=n_r+1}$, $u_k = [u_{1,k}, \ldots, u_{n_r,k}, 1]^T$, where $u_{i,k}$, $i = 1, \ldots, n_r$ are system inputs. For notational simplicity, let us define the following parameter vector:

$$p = \left[p^{(l)}(1)^T, \ldots, p^{(l)}(m)^T, p^{(n)}(1)^T, \ldots, p^{(n)}(n_h)^T \right]^T,$$

where $n_p = m(n_h + 1) + n_h(n_r + 1)$. Consequently, Eq. (3.1) can be written in a more compact form:

$$y_{M,k} = h(p, u_k), \tag{3.3}$$

where $h(\cdot)$ is a non-linear function representing the structure of a neural-network.

Let us assume that the system output satisfies the following equality:

$$y_k = y_{M,k} + v_k = h(p, u_k) + v_k, \tag{3.4}$$

where the noise v is zero-mean, Gaussian and uncorrelated in k, i.e., its statistics are

$$\mathcal{E}(v_k) = 0, \quad \mathcal{E}(v_i v_k^T) = \delta_{i,k} C, \tag{3.5}$$

where $C \in \mathbb{R}^{m \times m}$ is a known positive-definite matrix of the form $C = \sigma^2 I_m$, while σ^2 and $\delta_{i,k}$ stand for the variance and Kronecker's delta symbol, respectively. Under such an assumption, the theory of experimental design [7, 10, 11] can be exploited to develop a suitable training data set that allows obtaining a neural network with a considerably smaller uncertainty than those designed without it. First, let us define the so-called Fisher information matrix that constitutes a measure of parametric uncertainty of (3.1):

$$P^{-1} = \sum_{k=1}^{n_t} R_k R_k^T, \tag{3.6}$$

$$R_k = \left(\frac{\partial h(p, u_k)}{\partial p}\right)^T_{p = \hat{p}}, \tag{3.7}$$

and

$$\frac{\partial h(p, u_k)}{\partial p} = \begin{bmatrix} g\left(P^{(n)} u_k\right)^T & \mathbf{0}^T_{(m-1)(n_h+1)} & p_1^l(1)g'\left(u_k^T p^n(1)\right) u_k^T & \cdots \\ \vdots & \vdots & \vdots & \vdots \\ \mathbf{0}^T_{(m-1)(n_h+1)} & g\left(P^{(n)} u_k\right)^T & p_1^l(m)g'\left(u_k^T p^n(1)\right) u_k^T & \cdots \end{bmatrix}$$

$$\begin{matrix} \cdots & p_{n_h}^l(1)g'\left(u_k^T p^n(n_h)\right) u_k^T \\ & \vdots & \vdots \\ \cdots & p_{n_h}^l(m)g'\left(u_k^T p^n(n_h)\right) u_k^T \end{matrix} \Bigg], \tag{3.8}$$

where $g'(t) = \dfrac{dg(t)}{dt}$, $\hat{p}$ is a least-square estimate of p. It is easy to observe that the FIM (3.6) depends on the experimental conditions $\xi = [u_1, \ldots, u_{n_t}]$. Thus, optimal experimental conditions can be found by choosing u_k, $k = 1, \ldots, n_t$, so as to minimise some scalar function of (3.6). Such a function can be defined in several different ways [12, 13]:

• D-optimality criterion:

$$\Phi(\xi) = \det P, \tag{3.9}$$

• E-optimality criterion ($\lambda_{\max}(\cdot)$ stands for the maximum eigenvalue of its argument):

$$\Phi(\xi) = \lambda_{\max}(P); \tag{3.10}$$

• A-optimality criterion:

$$\Phi(\xi) = \text{trace } P; \tag{3.11}$$

- G-optimality criterion:

$$\Phi(\xi) = \max_{\boldsymbol{u}_k \in \mathbb{U}} \phi(\xi, \boldsymbol{u}_k), \tag{3.12}$$

where $\mathbb{U}$ stands for a set of admissible $\boldsymbol{u}_k$ that can be used for a system being considered (the design space), and

$$\phi(\xi, \boldsymbol{u}_k) = \text{trace}\left(\boldsymbol{R}_k^T \boldsymbol{P} \boldsymbol{R}_k\right) = \sum_{i=1}^m \boldsymbol{r}_{i,k} \boldsymbol{P} \boldsymbol{r}_{i,k}^T, \tag{3.13}$$

while $\boldsymbol{r}_{i,k}$ stands for the ith row of $\boldsymbol{R}_k^T$;
- Q-optimality criterion:

$$\Phi(\xi) = \text{trace}\left(\boldsymbol{P}_Q^{-1} \boldsymbol{P}\right), \tag{3.14}$$

where $\boldsymbol{P}_Q^{-1} = \int \boldsymbol{R}(\boldsymbol{u}) \boldsymbol{R}(\boldsymbol{u})^T \mathrm{d}Q(\boldsymbol{u})$, $\boldsymbol{R}(\boldsymbol{u}) = \left(\frac{\partial h(\boldsymbol{p},\boldsymbol{u})}{\partial \boldsymbol{p}}\right)^T_{\boldsymbol{p}=\hat{\boldsymbol{p}}}$, and Q stands for the so-called environmental probability, which gives independent input vectors in the actual environment where a trained network is to be exploited [12].

As has already been mentioned, a valuable property of the FIM is that its inverse constitutes an approximation of the covariance matrix for $\hat{\boldsymbol{p}}$ [14], i.e., it is a lower bound of this covariance matrix that is established by the so-called Cramér–Rao inequality [14]:

$$\text{cov}(\hat{\boldsymbol{p}}) \succeq \boldsymbol{P}. \tag{3.15}$$

Thus, a D-optimum design minimises the volume of the confidence ellipsoid approximating the feasible parameter set of (3.1) (see, e.g., [10] for further explanations). An E-optimum design minimises the length of the largest axis of the same ellipsoid. An A-optimum design suppresses the average variance of parameter estimates. A G-optimum design minimises the variance of the estimated response of (3.1). Finally, a Q-optimum design minimises the expectation of the generalisation error $\mathcal{E}(\varepsilon_{\text{gen}})$ defined by [12]:

$$\varepsilon_{\text{gen}} = \int \|h(\boldsymbol{p}, \boldsymbol{u}) - h(\hat{\boldsymbol{p}}, \boldsymbol{u})\|^2 \mathrm{d}Q(\boldsymbol{u}). \tag{3.16}$$

Among the above-listed optimality criteria, the D-optimality criterion, due to its simple updating formula, has been employed by many authors in the development of computer algorithms for calculating optimal experimental design. Another important property is that D-optimum design is invariant to non-degenerate linear transformation of the model. This property is to be exploited and suitably discussed in Sect. 3.1.2. It is also important to underline that, from the practical point of view, D-optimum designs often perform well according to other criteria (see [10] and the references therein for more details). For further explanations regarding D-optimality criteria, the

Fig. 3.1 ith output of the system and its bounds obtained with a neural network

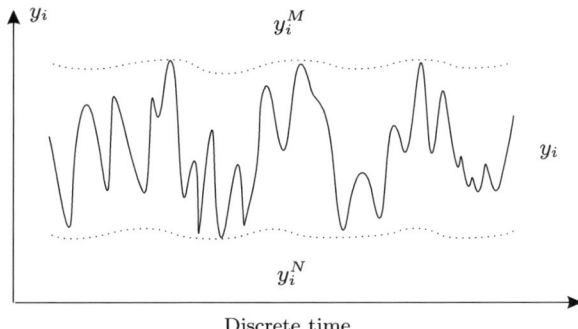

reader is referred to the excellent textbooks [7, 10, 13, 15]. Since the research results presented in this section are motivated by fault diagnosis applications of neural networks, the main objective is to use such a design criterion which makes it possible to obtain accurate bounds of the system output (cf. Fig. 3.1). Indeed, it is rather vain to assume that it is possible to develop a neural network with an arbitrarily small uncertainty, i.e., to obtain a perfect model of the system. A more realistic task is to design a model that will provide a reliable knowledge about the bounds of the system output that reflect the expected behaviour of the system. This is especially important from the point of view of robust fault diagnosis. The design methodology of such robust techniques relies on the idea that fault diagnosis and control schemes should perform reliably for any system behaviour that is consistent with output bounds. This is in contradiction with the conventional approaches, where fault diagnosis and control schemes are designed to be optimal for one single model. The bounds presented in Fig. 3.1 can be described as follows:

$$ y_{i,k}^N \leq y_{i,k} \leq y_{i,k}^M, \quad i = 1, \ldots, m. \tag{3.17} $$

In [16], the authors developed an approach that can be used for determining (3.17) (that forms the $100(1 - \alpha)$ confidence interval of $y_{i,k}$) for single output ($m = 1$) neural networks. In [3], the approach of [16] was extended to multi-output models. If the neural model gives a good prediction of the actual system behaviour, then $\hat{p}$ is close to the optimal parameter vector and the following first-order Taylor expansion of (3.4) can be exploited:

$$ \mathbf{y}_k \approx \hat{\mathbf{y}}_k + \mathbf{R}_k^T (\mathbf{p} - \hat{\mathbf{p}}) + \mathbf{v}_k, \quad \hat{\mathbf{y}}_k = \mathbf{h}\left(\hat{\mathbf{p}}, \mathbf{u}_k\right). \tag{3.18} $$

Thus, assuming that $\mathcal{E}(\hat{\mathbf{p}}) = \mathbf{p}$, we get

$$ \mathcal{E}(\mathbf{y}_k - \hat{\mathbf{y}}_k) \approx \mathbf{R}_k^T (\mathbf{p} - \mathcal{E}(\hat{\mathbf{p}})) + \mathcal{E}(\mathbf{v}_k) \approx \mathbf{0}. \tag{3.19} $$

Using a similar approach, the covariance matrix is given by

$$\text{cov}(\boldsymbol{y}_k - \hat{\boldsymbol{y}}_k) = \mathcal{E}\left((\boldsymbol{y}_k - \hat{\boldsymbol{y}}_k)(\boldsymbol{y}_k - \hat{\boldsymbol{y}}_k)^T\right)$$

$$\approx \boldsymbol{R}_k^T \mathcal{E}\left((\boldsymbol{p} - \hat{\boldsymbol{p}})(\boldsymbol{p} - \hat{\boldsymbol{p}})^T\right) \boldsymbol{R}_k + \sigma^2 \boldsymbol{I}_m. \tag{3.20}$$

Using the classic results regarding $\mathcal{E}\left((\boldsymbol{p} - \hat{\boldsymbol{p}})(\boldsymbol{p} - \hat{\boldsymbol{p}})^T\right)$ [7, 10, 13], i.e.,

$$\mathcal{E}\left((\boldsymbol{p} - \hat{\boldsymbol{p}})(\boldsymbol{p} - \hat{\boldsymbol{p}})^T\right) = \sigma^2 \boldsymbol{P}, \tag{3.21}$$

Equation (3.20) can be expressed as

$$\text{cov}(\boldsymbol{y}_k - \hat{\boldsymbol{y}}_k) \approx \sigma^2 \left(\boldsymbol{R}_k^T \boldsymbol{P} \boldsymbol{R}_k + \boldsymbol{I}_m\right). \tag{3.22}$$

Subsequently, using (3.22), the standard deviation of $y_{i,k} - \hat{y}_{i,k}$ is given by

$$\sigma_{y_{i,k} - \hat{y}_{i,k}} = \sigma\left(1 + \boldsymbol{r}_{i,k} \boldsymbol{P} \boldsymbol{r}_{i,k}^T\right)^{1/2}, \quad i = 1, \ldots, m. \tag{3.23}$$

Using (3.23) and the result of [16], it can be shown that $y_{i,k}^N$ and $y_{i,k}^M$ (that form the $100(1 - \alpha)$ confidence interval of $y_{i,k}$) can be approximated as follows:

$$y_{i,k}^N = \hat{y}_{i,k} - t_{n_t - n_p}^{\alpha/2} \hat{\sigma}\left(1 + \boldsymbol{r}_{i,k} \boldsymbol{P} \boldsymbol{r}_{i,k}^T\right)^{1/2}, \quad i = 1, \ldots, m, \tag{3.24}$$

$$y_{i,k}^M = \hat{y}_{i,k} + t_{n_t - n_p}^{\alpha/2} \hat{\sigma}\left(1 + \boldsymbol{r}_{i,k} \boldsymbol{P} \boldsymbol{r}_{i,k}^T\right)^{1/2}, \quad i = 1, \ldots, m, \tag{3.25}$$

where $t_{n_t - n_p}^{\alpha/2}$ is the t-Student distribution quantile, and $\hat{\sigma}$ is the standard deviation estimate. Bearing in mind the fact that the primary purpose is to develop reliable bounds of the system output, it is clear from (3.17), (3.24), and (3.25) that the G-optimality criterion should be selected.

When some experiments are repeated, then the number n_e of distinct $\boldsymbol{u}_k$s is smaller than the total number of observations n_t. The design resulting from this approach is called continuous experimental design and it can be described as follows:

$$\xi = \left\{ \begin{matrix} \boldsymbol{u}_1 & \boldsymbol{u}_2 & \ldots & \boldsymbol{u}_{n_e} \\ \mu_1 & \mu_2 & \ldots & \mu_{n_e} \end{matrix} \right\}, \tag{3.26}$$

where $\boldsymbol{u}_k$s are said to be the *support points*, and $\mu_1, \ldots, \mu_{n_e}, \mu_k \in [0, 1]$ are called their weights, which satisfy $\sum_{k=1}^{n_e} \mu_k = 1$. Thus, when the design (3.26) is optimal (with respect to one of the above-defined criteria), then the support points can also be called *optimal inputs*. Thus, the Fisher information matrix can now be defined as follows:

$$\boldsymbol{P}^{-1} = \sum_{k=1}^{n_e} \mu_k \boldsymbol{R}_k \boldsymbol{R}_k^T. \tag{3.27}$$

The fundamental property of continuous experimental design is the fact that the optimum designs resulting from the D-optimality and G-optimality criteria are the same (the Kiefer–Wolfowitz equivalence theorem [7, 10, 13]). Another reason for using D-optimum design is the fact that it is probably the most popular criterion. Indeed, most of the algorithms presented in the literature are developed for D-optimum design. Bearing in mind all of the above-mentioned circumstances, the subsequent part of this section is devoted to D-optimum experimental design. The next section shows an illustrative example whose results clearly show profits that can be gained while applying D-OED to neural networks.

3.1.1 Illustrative Example

Let us consider a neuron model with the logistic activation function [3]:

$$y_{M,k} = \frac{p_1}{1 + e^{-p_2 u_k - p_3}}. \tag{3.28}$$

It is obvious that the continuous experimental design for the model (3.28) should have at least three different support points ($n_p = 3$ for (3.28)). For a three-point design, the determinant of the FIM (3.27) is

$$\det \boldsymbol{P}^{-1} = \frac{p_1^4}{p_2^2} \mu_1 \mu_2 \mu_3 e^{2x_1} e^{2x_2} e^{2x_3}$$
$$\cdot \frac{((e^{x_2} - e^{x_1})x_3 + (e^{x_3} - e^{x_2})x_1 + (e^{x_1} - e^{x_3})x_2)^2}{(e^{x_1} + 1)^4(e^{x_2} + 1)^4(e^{x_3} + 1)^4}, \tag{3.29}$$

where $x_i = p_2 u_i + p_3$. Bearing in mind the fact that the minimisation of (3.9) is equivalent to the maximisation of (3.29), a numerical solution regarding the D-optimum continuous experimental design can be written as

$$\xi = \left\{ \begin{matrix} \boldsymbol{u}_1 \ \boldsymbol{u}_2 \ \boldsymbol{u}_3 \\ \mu_1 \ \mu_2 \ \mu_3 \end{matrix} \right\}$$
$$= \left\{ \begin{pmatrix} \frac{1.041 - p_3}{p_2}, 1 \end{pmatrix} \begin{pmatrix} \frac{-1.041 - p_3}{p_2}, 1 \end{pmatrix} \begin{pmatrix} \frac{x_3 - p_3}{p_2}, 1 \end{pmatrix} \\ \frac{1}{3} \qquad\qquad \frac{1}{3} \qquad\qquad \frac{1}{3} \end{pmatrix} \right\}, \tag{3.30}$$

whereas x_3 is an arbitrary constant satisfying $x_3 \geq \zeta, \zeta \approx 12$. In order to check if the design (3.30) is really D-optimum, the Kiefer–Wolfowitz equivalence theorem [7, 10] can be employed. In the light of this theorem, the design (3.30) is D-optimum when

Fig. 3.2 Variance function for (3.30) and $x_3 = 20$

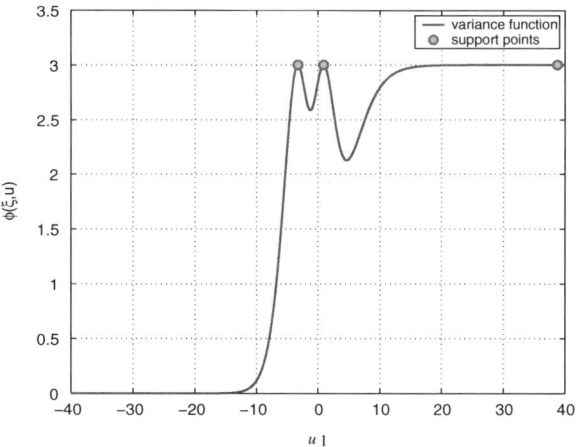

$$\phi(\xi, \boldsymbol{u}_k) = \text{trace}\left(\boldsymbol{R}_k^T \boldsymbol{P} \boldsymbol{R}_k\right) \leq n_p, \tag{3.31}$$

where the equality holds for measurements described by (3.30). It can be seen from Fig. 3.2 that the design (3.30) satisfies (3.31). This figure also justifies the role of the constant ζ, which is a lower bound of x_3 in the third support point of (3.30). Indeed, it can be observed that the design (3.30) is D-optimum (the variance function is $n_p = 3$) when the third support point is larger than some constant value, which is equivalent to $x_3 \geq \zeta$.

In order to justify the effectiveness of (3.30), let us assume that the nominal parameter vector is $\boldsymbol{p} = [2, 0.5, 0.6]^T$. It is also assumed that $n_t = 9$. This means that each of the measurements consistent with (3.30) should be repeated 3 times. For the purpose of comparison, a set of n_t points was generated according to the uniform distribution $\mathcal{U}(-4, 40)$. It should also be pointed out that v was generated according to $\mathcal{N}(0, 0.1^2)$. Figure 3.3 presents feasible parameter sets obtained with the strategies considered. These sets are defined according to the following formula [7]:

$$\mathbb{P} = \left\{ \boldsymbol{p} \in \mathbb{R}^{n_p} \,\Big|\, \sum_{k=1}^{n_t} (y_k - f(\boldsymbol{p}, \boldsymbol{u}_k))^2 \leq \sigma^2 \chi^2_{\alpha, n_t} \right\}, \tag{3.32}$$

where χ^2_{α, n_t} is the Chi-square distribution quantile. From Fig. 3.3, it is clear that the application of D-OED results in a model with a considerably smaller uncertainty than the one designed without it. These results also imply that the system bounds (3.17) will be more accurate.

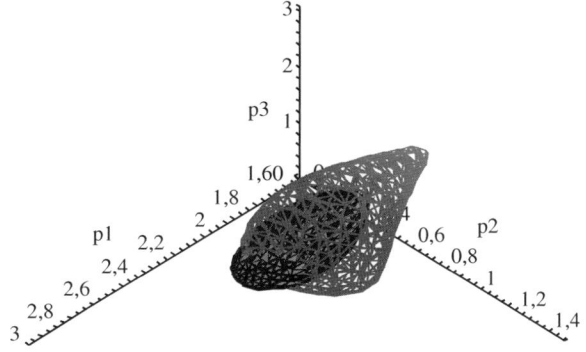

Fig. 3.3 Feasible parameter set obtained for (3.30) (smaller) and for a set of randomly generated points

3.1.2 Algorithms and Properties of D-OED for Neural Networks

Regularity of the FIM

The Fisher information matrix $\boldsymbol{P}^{-1}$ of (3.1) may be singular for some parameter configurations, and in such cases it is impossible to obtain its inverse $\boldsymbol{P}$ necessary to calculate (3.17), as well as to utilise specialised algorithms for obtaining D-optimum experimental design [7, 10]. In [17] the author established the conditions under which $\boldsymbol{P}^{-1}$ is singular. These conditions can be formulated as follows

Theorem 3.1 *[17] The Fisher information matrix $\boldsymbol{P}^{-1}$ of (3.1) is singular iff at least one of the following conditions holds true:*

1. *There exists j such that $[p_{j,1}^{(n)}, \ldots, p_{j,n_r}^{(n)}]^T = \boldsymbol{0}$.*
2. *There exists j such that $[p_{1,j}^{(l)}, \ldots, p_{m,j}^{(l)}]^T = \boldsymbol{0}$.*
3. *There exist different j_1 and j_2 such that $\boldsymbol{p}^{(n)}(j_1) = \pm \boldsymbol{p}^{(n)}(j_2)$.*

A direct consequence of the above theorem is that a network with singular $\boldsymbol{P}^{-1}$ can be reduced to one with positive definite $\boldsymbol{P}^{-1}$ by removing redundant hidden neurons. Based on this property, it is possible to develop a procedure that can be used for removing the redundant neurons without performing the retraining of a network [18].

If the conditions of Theorem 3.1 indicate that $\boldsymbol{P}^{-1}$ is not singular, then the strategy of collecting measurements according to the theory of D-optimum experimental design (the maximisation of the determinant of $\boldsymbol{P}^{-1}$) guarantees that the Fisher information matrix is positive definite. This permits approximating an exact feasible parameter set (3.32) with an ellipsoid (cf. Fig. 3.3 to see the similarity to an ellipsoid). Unfortunately, the conditions of Theorem 3.1 have strictly theoretical meaning as in most practical situations the FIM would be close to singular but not singular in an exact sense. This makes the process of eliminating redundant hidden neurons far more difficult, and there is no really efficient algorithm that could be employed to settle this problem. Indeed, the approach presented in [12] is merely sub-optimal.

On the other hand, if such an algorithm does not give satisfactory results, i.e., the FIM is still close to the singular matrix, then the FIM should be regularised in the following way [10, p. 110]:

$$P_\kappa^{-1} = P^{-1} + \kappa I, \tag{3.33}$$

for $\kappa > 0$ small but large enough to permit the inversion of P_κ^{-1}.

Relation Between Non-Linear Parameters and D-OED

Dependence on parameters that enter non-linearly ((3.30) depends on p_2 and p_3 but does not depend on p_1) into the model is an unappealing characteristic of non-linear optimum experimental design. As has already been mentioned, there is a number of works dealing with D-OED for neural networks but none of them has exploited this important property. In [3], it was shown that experimental design for a general structure (3.1) is independent of parameters that enter linearly into (3.1). Indeed, it can be shown that (3.8) can be transformed into an equivalent form:

$$\frac{\partial h\,(p, u_k)}{\partial p} = L\left(P^{(n)}, u_k\right) Z\left(P^{(l)}\right), \tag{3.34}$$

and

$$
L\left(P^{(n)}, u_k\right)
= \begin{bmatrix}
g\left(P^{(n)}u_k\right)^T & 0_{(m-1)(n_h+1)} & \left(g'\left(P^{(n)}u_k\right) \otimes u_k\right)^T \\
\vdots & \vdots & \vdots \\
0_{(m-1)(n_h+1)} & g\left(P^{(n)}u_k\right)^T & \left(g'\left(P^{(n)}u_k\right) \otimes u_k\right)^T
\end{bmatrix}, \tag{3.35}
$$

$$
Z\left(P^{(l)}\right) = \begin{bmatrix}
1_{n_h+1} & \cdots & x_{(m-1)(n_h+1)} \\
x_{(m-1)(n_h+1)} & \cdots & 1_{n_h+1} \\
p_1^{(l)}(1)1_{n_r+1} & \cdots & p_1^{(l)}(m)1_{n_r+1} \\
\vdots & \vdots & \vdots \\
p_{n_h+1}^{(l)}(1)1_{n_r+1} & \cdots & p_{n_h+1}^{(l)}(m)1_{n_r+1}
\end{bmatrix}, \tag{3.36}
$$

where $\otimes$ denotes the Kronecker product, x_t stands for an arbitrary t-dimensional vector, and

$$g'(t) = [g'(t_1), \dots, g'(t_{n_h})]^T. \tag{3.37}$$

Thus, R_k can be written in the following form:

$$R_k = P_1 R_{1,k}, \tag{3.38}$$

where

$$P_1 = \left(Z\left(P^{(l)} \right) \right)^T_{p=\hat{p}} \tag{3.39}$$

and

$$R_{1,k} = \left(L\left(P^{(n)}, u_k \right) \right)^T_{p=\hat{p}}. \tag{3.40}$$

The Fisher information matrix is now given by

$$P^{-1} = \sum_{k=1}^{n_t} R_k R_k^T = P_1 \left[\sum_{k=1}^{n_t} R_{1,k} R_{1,k}^T \right] P_1^T. \tag{3.41}$$

Thus, the determinant of P^{-1} is given by

$$\det\left(P^{-1} \right) = \det\left(P_1 \right)^2 \det\left(\sum_{k=1}^{n_t} R_{1,k} R_{1,k}^T \right). \tag{3.42}$$

From (3.42) it is clear that the process of minimising the determinant of P^{-1} with respect to u_ks is independent of the linear parameters p^l. This means that at least a rough estimate of $P^{(n)}$ is required to solve the experimental design problem. Such estimates can be obtained with any training method for feed-forward neural networks [19]. A particularly interesting approach was recently developed in [20]. The authors proposed a novel method of backpropagating the desired response through the layers of the MLP in such a way as to minimise the mean-square error. Thus, the obtained solution may constitute a good starting point for experimental design.

Indeed, it is rather vain to expect that it is possible to obtain a design that is to be appropriate for all networks of a given structure. It is very easy to imagine two neural networks of the same structure that may represent two completely different systems. If some rough estimates are given, then specialised algorithms for D-optimum experimental design can be applied [7, 10].

σ-Equivalence Theorem for D-OED

Undoubtedly, the most popular activation functions $g(\cdot)$ that are commonly employed for designing neural networks are $g_\sigma(t) = \frac{1}{1+\exp(-t)}$ and $g_{tg}(t) = \tanh(t)$. It is well known that these functions are very similar, and this similarity is expressed by the following relationship:

$$g_\sigma(t) = \frac{1}{2} + \frac{1}{2} g_{tg}\left(\frac{1}{2}t \right). \tag{3.43}$$

Thus, the problem is to show how to use a D-optimum design obtained for a network with the activation functions $g_\sigma(\cdot)$ to obtain a D-optimum design for a network with the activation functions $g_{\text{tg}}(\cdot)$. In this work, the above problem is solved as follows

Theorem 3.2 *Let*

$$\xi_\sigma = \left\{ \begin{array}{ccc} \boldsymbol{u}_1 & \ldots & \boldsymbol{u}_{n_e} \\ \mu_1 & \ldots & \mu_{n_e} \end{array} \right\} \tag{3.44}$$

denote a D-optimum design for the network

$$\boldsymbol{y}_{M,k} = \boldsymbol{P}^{(l)} g_{\text{tg}} \left(\boldsymbol{P}^{(n)} \boldsymbol{u}_k \right). \tag{3.45}$$

Then the design (3.44) is D-optimum for the following network:

$$\boldsymbol{y}_{M,k} = \boldsymbol{P}_\sigma^{(l)} g_\sigma \left(\boldsymbol{P}^{(n)} \boldsymbol{u}_k^\sigma \right), \tag{3.46}$$

where $\boldsymbol{u}_k^\sigma = 2\boldsymbol{u}_k$ and $\boldsymbol{P}_\sigma^{(l)}$ is an arbitrary (non-zero) matrix.

Proof It is straightforward to observe that

$$\boldsymbol{P}_\sigma^{(l)} g_\sigma \left(2\boldsymbol{P}^{(n)} \boldsymbol{u}_k \right) = \boldsymbol{P}_{\text{tg}}^{(l)} g_{\text{tg}} \left(\boldsymbol{P}^{(n)} \boldsymbol{u}_k \right), \tag{3.47}$$

where

$$\boldsymbol{P}_{\text{tg}}^{(l)} = \frac{1}{2} \boldsymbol{P}_\sigma^{(l)} + \left[\boldsymbol{0}_{m \times n_h}, \frac{1}{2} \boldsymbol{P}_\sigma^{(l)} \boldsymbol{1}_{n_h+1} \right]. \tag{3.48}$$

Thus, using (3.42), the determinant of the FIM for (3.45) is

$$\det \left(\boldsymbol{P}^{-1} \right) = \det \left(\left(\boldsymbol{Z} \left(\boldsymbol{P}^{(l)} \right) \right)_{p=\hat{p}}^T \right)^2 \det \left(\sum_{k=1}^{n_t} \boldsymbol{R}_{1,k} \boldsymbol{R}_{1,k}^T \right), \tag{3.49}$$

while using (3.42), (3.47), and (3.48), the determinant of the Fisher information matrix for (3.46) is

$$\det \left(\boldsymbol{P}^{-1} \right) = \det \left(\left(\boldsymbol{Z} \left(\boldsymbol{P}_{\text{tg}}^{(l)} \right) \right)_{p=\hat{p}}^T \right)^2 \det \left(\sum_{k=1}^{n_t} \boldsymbol{R}_{1,k} \boldsymbol{R}_{1,k}^T \right), \tag{3.50}$$

and $\boldsymbol{R}_{1,k}$ in (3.49) and (3.50) is calculated by substituting

$$L\left(\boldsymbol{P}^{(n)}, \boldsymbol{u}_k\right)$$

$$= \begin{bmatrix} \boldsymbol{g}_{\mathrm{tg}}\left(\boldsymbol{P}^{(n)}\boldsymbol{u}_k\right)^T & \boldsymbol{0}_{(m-1)(n_h+1)} & \left(\boldsymbol{g}'_{\mathrm{tg}}\left(\boldsymbol{P}^{(n)}\boldsymbol{u}_k\right) \otimes \boldsymbol{u}_k\right)^T \\ \vdots & \vdots & \vdots \\ \boldsymbol{0}_{(m-1)(n_h+1)} & \boldsymbol{g}_{\mathrm{tg}}\left(\boldsymbol{P}^{(n)}\boldsymbol{u}_k\right)^T & \left(\boldsymbol{g}'_{\mathrm{tg}}\left(\boldsymbol{P}^{(n)}\boldsymbol{u}_k\right) \otimes \boldsymbol{u}_k\right)^T \end{bmatrix} \tag{3.51}$$

and

$$\boldsymbol{g}'_{\mathrm{tg}}(t) = \left[g'_{\mathrm{tg}}(t_1), \ldots, g'_{\mathrm{tg}}(t_{n_h})\right]^T \tag{3.52}$$

into (3.40).

From (3.49) and (3.50), it is clear that the D-optimum design obtained with either (3.49) or (3.50) is identical, which completes the proof.

Based on the above results, the following remark can be formulated:

Remark 3.3 Theorem 3.2 and the Kiefer–Wolfowitz theorem [7, 10] imply that the D-optimum design (3.44) for (3.45) is also G-optimum for this model structure and, hence, it is G-optimum for (3.46).

Illustrative Example

Let us reconsider the example presented in Sect. 3.1.1. The purpose of this example was to obtain a D-optimum experimental design for the model (3.28). As a result, the design (3.30) was determined. The purpose of further deliberations is to apply the design (3.30) to the following model:

$$y_{M,k} = p_1 \tanh(2(p_2 u_k + p_3)), \tag{3.53}$$

and to check if it is D-optimum for (3.53). Figure 3.4 presents the variance function (3.12) for the model (3.53). From this figure it is clear that the variance function satisfies the D-optimality condition (3.31).

Wynn–Fedorov Algorithm for the MLP

The preceding part of Sect. 3.1.2 presents important properties of D-OED for neural networks. In this section, these properties are exploited to develop an effective algorithm for calculating D-OED for neural networks. In a numerical example of Sect. 3.1.1, it is shown how to calculate D-OED for a neural network composed of one neuron only. In particular, the algorithm was reduced to direct optimisation of the determinant of the FIM with respect to experimental conditions. This means that non-linear programming techniques have to be employed to settle this

Fig. 3.4 Variance function
for (3.30) ($x_3 = 20$) and
model (3.53)

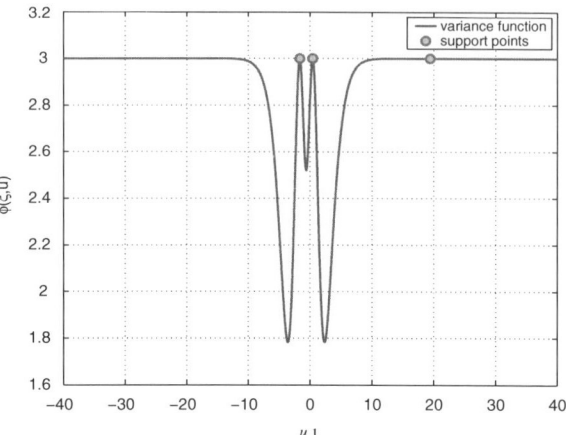

problem. Unfortunately, it should be strongly underlined that such an approach is
impractical when larger neural networks are investigated. Fortunately, the Kiefer–
Wolfowitz equivalence theorem [7, 10, 13] (see also (3.31)) provides some guidance
useful in construction of a suitable numerical algorithm. The underlying reasoning
boils down to correcting a non-optimum design ξ_k (obtained after k iterations) by a
convex combination with another design [15, p. 27], which hopefully improves the
current solution, i.e.,

$$\xi_{k+1} = (1 - \alpha_k)\xi_k + \alpha_k \xi(\boldsymbol{u}_k) \tag{3.54}$$

for some convenient $0 < \alpha_k < 1$, where

$$\xi_k = \left\{ \begin{matrix} \boldsymbol{u}_1\ \boldsymbol{u}_2\ \dots\ \boldsymbol{u}_{n_e} \\ \mu_1\ \mu_2\ \dots\ \mu_{n_e} \end{matrix} \right\}, \quad \xi(\boldsymbol{u}_k) = \left\{ \begin{matrix} \boldsymbol{u}_k \\ 1 \end{matrix} \right\}, \tag{3.55}$$

while the convex combination (3.54) is realised as follows:
If $\boldsymbol{u}_k \neq \boldsymbol{u}_i$, $i = 1, \dots, n_e$, then

$$\xi_{k+1} = \left\{ \begin{matrix} \boldsymbol{u}_1 & \boldsymbol{u}_2 & \dots & \boldsymbol{u}_{n_e} & \boldsymbol{u}_k \\ (1 - \alpha_k)\mu_1 & (1 - \alpha_k)\mu_2 & \dots & (1 - \alpha_k)\mu_{n_e} & \alpha_k \end{matrix} \right\}, \tag{3.56}$$

else

$$\xi_{k+1} = \left\{ \begin{matrix} \boldsymbol{u}_1 & \boldsymbol{u}_2 & \dots & \boldsymbol{u}_i & \dots & \boldsymbol{u}_{n_e} \\ (1 - \alpha_k)\mu_1 & (1 - \alpha_k)\mu_2 & \dots & (1 - \alpha_k)\mu_i + \alpha_k & \dots & (1 - \alpha_k)\mu_{n_e} \end{matrix} \right\}.$$

In this way, the experimental effort related to ξ_k is reduced and measurements corresponding to $\xi(\boldsymbol{u}_k)$ are favored instead. Hence, the problem is to select $\xi(\boldsymbol{u}_k)$ so
as to get a better value of the optimality criterion. A solution to this problem can
be found with the help of the Kiefer–Wolfowitz equivalence theorem [7, 10, 13].

Indeed, the support points of optimum design ξ^* coincide with the maxima of the variance function $\phi(\xi^*, \boldsymbol{u}_k)$ (see Sect. 3.1.1 for an illustrative example). Thus, by the addition of $\xi(\boldsymbol{u}_k)$ for which the maximum of $\phi(\xi_k, \boldsymbol{u}_k)$ is attained, an improvement in the current design can be expected (see [7, 13] for more details).

The above-outlined approach forms the base of the celebrated Wynn–Fedorov algorithm [7, 10]. In this section, the results developed in Sect. (3.1.2) and the one of Theorem 3.1 are utilised to adapt the Wynn–Fedorov algorithm in order to develop D-OED for neural networks.

First, let us start with a slight modification of the Wynn–Fedorov algorithm that boils down to reducing the necessity of using the linear parameters of (3.1) in the computational procedure.

Since, according to (3.38),

$$R_k = P_1 R_{1,k} \, , \tag{3.57}$$

then (3.12) can be written as follows:

$$\Phi(\xi) = \max_{\boldsymbol{u}_k \in \mathbb{U}} \text{trace} \left(R_{1,k}^T P_1^T P P_1 R_{1,k} \right) . \tag{3.58}$$

Using (3.41) and the notation of continuous experimental design, the matrix P can be expressed as follows:

$$P = \left(P_1^T \right)^{-1} \left[\sum_{k=1}^{n_e} \mu_k R_{1,k} R_{1,k}^T \right]^{-1} P_1^{-1} . \tag{3.59}$$

It is easy to observe that if Condition 2 of Theorem 3.1 is not satisfied, then it is possible to compute the inverse of P_1 (cf. (3.39)). Similarly, if both Conditions 1 and 3 are not satisfied, then the matrix

$$P_2 = \left[\sum_{k=1}^{n_e} \mu_k R_{1,k} R_{1,k}^T \right]^{-1} \tag{3.60}$$

in (3.59) can be calculated. Now, (3.58) can be expressed in the following form:

$$\Phi(\xi) = \max_{\boldsymbol{u}_k \in \mathbb{U}} \phi(\xi, \boldsymbol{u}_k), \tag{3.61}$$

where

$$\phi(\xi, \boldsymbol{u}_k) = \text{trace} \left(R_{1,k}^T P_2 R_{1,k} \right) . \tag{3.62}$$

Note that the computation of (3.61) does not require any knowledge about the parameter matrix $P^{(l)}$, which enters linearly into (3.1). Another advantage is that it is not necessary to form the matrix $P_1 \in \mathbb{R}^{n_p \times n_p}$, which then has to be multiplied by $R_{1,k}$ to form (3.57). This implies a reduction in the computational burden.

Equation (3.61) can be perceived as the main step of the Wynn–Fedorov algorithm, which can now be described as follows:

Step 0: Obtain an initial estimate of $\boldsymbol{P}^{(n)}$, i.e., the parameter matrix that enters non-linearly into (3.1), with any method for training the MLP [19]. Set $k = 1$, choose a non-degenerate $(\det(\boldsymbol{P}^{-1}) \neq 0$, it is satisfied when no conditions of Theorem 3.1 are fulfilled) design ξ_k, set the maximum number of iterations $n_{\max}$.
Step 1: Calculate

$$\boldsymbol{u}_k = \arg \max_{\boldsymbol{u}_k \in \mathbb{U}} \text{trace} \left(\boldsymbol{R}_{1,k}^T \boldsymbol{P}_2 \boldsymbol{R}_{1,k} \right). \tag{3.63}$$

Step 2: If $\phi(\xi_k, \boldsymbol{u}_k)/n_p < 1 + \epsilon$, where $\epsilon > 0$, is sufficiently small, then STOP, else go to *Step 3*.
Step 3: Calculate a weight associated with a new support point $\boldsymbol{u}_k$ according to

$$\alpha_k = \arg \max_{0<\alpha<1} \det \left((1 - \alpha) \boldsymbol{P}_2 + \alpha \boldsymbol{R}_{1,k} \boldsymbol{R}_{1,k}^T \right), \tag{3.64}$$

which for single-output systems $(m = 1)$ is given by

$$\alpha_k = \frac{\phi(\xi_k, \boldsymbol{u}_k) - n_p}{(\phi(\xi_k, \boldsymbol{u}_k) - 1)n_p}, \tag{3.65}$$

and go to *Step 4*.
Step 4: Obtain a new design ξ_{k+1}, which is a convex combination [15, p. 27] of the form

$$\xi_{k+1} = (1 - \alpha_k)\xi_k + \alpha_k \xi(\boldsymbol{u}_k). \tag{3.66}$$

If $k = n_{\max}$, then STOP, else set $k = k + 1$ and go to *Step 1*.

Step 1 is crucial in the presented algorithm. Indeed, the first problem is the fact that the calculation of (3.60) involves matrix inversions. Since the dimension of this matrix equals n_p, then, even for simple networks, the number of parameters is a dozen or so. Subsequently, it is shown that effective recursive formulae can be established for calculating $\boldsymbol{P}_2^k$, i.e., the matrix $\boldsymbol{P}_2$ in the kth iteration of the Wynn–Fedorov algorithm. It can be seen from the inverse of (3.60) and (3.56) that

$$\left(\boldsymbol{P}_2^{k+1} \right)^{-1} = (1 - \alpha_k) \left(\boldsymbol{P}_2^k \right)^{-1} + \alpha_k \boldsymbol{R}_{1,k} \boldsymbol{R}_{1,k}^T. \tag{3.67}$$

Using the matrix inversion lemma and (3.67), the following recursive relation can be established:

$$P_2^{k+1} = \frac{1}{1-\alpha_k}$$

$$\cdot \left[P_2^k - P_2^k R_{1,k} \left[\frac{1-\alpha_k}{\alpha_k} I_m + R_{1,k}^T P_2^k R_{1,k} \right]^{-1} R_{1,k}^T P_2^k \right]. \qquad (3.68)$$

Note that the calculation of (3.68) requires inversion of an m-dimensional matrix instead of an n_p-dimensional one.

The second problem concerning Step 1 is the fact that the variance function (3.63) is multi-modal and hence conventional optimisation routines cannot be applied to settle (3.61). For further explanations concerning the problem (3.61), the reader is referred to [21]. Based on numerous computer experiments, it has been found that the extremely simple Adaptive Random Search (ARS) algorithm [7] is especially well suited for the purpose of optimising (3.61), although other techniques such as evolutionary algorithms [22] can successfully be applied as well.

It is important to note that the above algorithm makes use of information about the gradient of the performance index only, and the rule (3.64) results in the steepest-descent algorithm. As a result, the convergence rate of the algorithm is comparable with its gradient counterparts from mathematical non-linear programming. This implies a significant decrease in the performance index in the first few iterations, but then serious moderation of the convergence rate occurs as the optimum is approached. There are some second-order counterparts of the algorithm considered, but they require significantly higher implementation complexity. However, it should be pointed out that they may improve the design weight rather than the support points, and in this context the features of the presented algorithm are satisfactory. Indeed, many computer experiments show that the most significant support points are found in just several iterations.

Numerical Example

The problem is to approximate the function

$$y_k = \exp(-\sin(u_k)) + v_k,$$

where $v \sim \mathcal{N}(0, 0.02^2)$, $u_k \in [0.1, 10]$, with a neural network containing $n_h = 4$ hidden neurons with hyperbolic tangent activation functions. Thus, the number of parameters to be estimated is $n_p = 13$. In the preliminary experiment $u_k, k = 1, \ldots, n_t = 15$ were obtained in such a way as to equally divide the design space $\mathbb{U} \in [0.1, 10]$. Then the Levenberg–Marquardt algorithm [7] was employed for parameter estimation. Based on the obtained parameter estimates, the Wynn–Fedorov algorithm was utilised to obtain D-OED, and then the parameter estimation process was repeated once again. Figure 3.5 shows the variance function and D-optimum inputs (support points). Note that the number of support points is n_p while $\mu_k = 1/13$, $k = 1, \ldots, 13$. Based on the obtained design, $n_t = 13$ measurements were

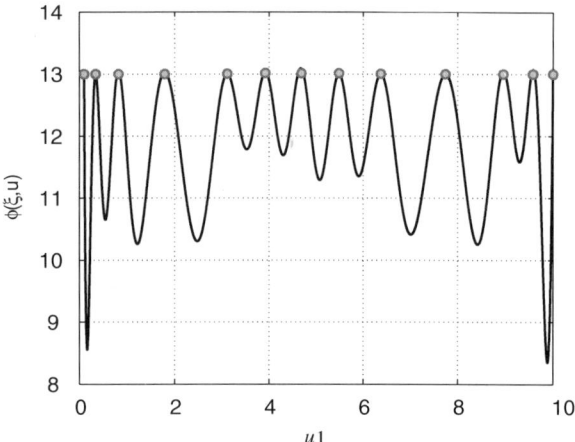

Fig. 3.5 Variance function and the corresponding support points

taken, each corresponding to the subsequent support points. Figure 3.6 presents the
output bounds (3.17) for the network obtained with the application of OED (the 2nd
net) and the one obtained without it (the 1st net), while the true output represents the
shape of the approximated function. It can be observed that the use of OED results
in a network with a significantly smaller uncertainty than the one designed without
it.

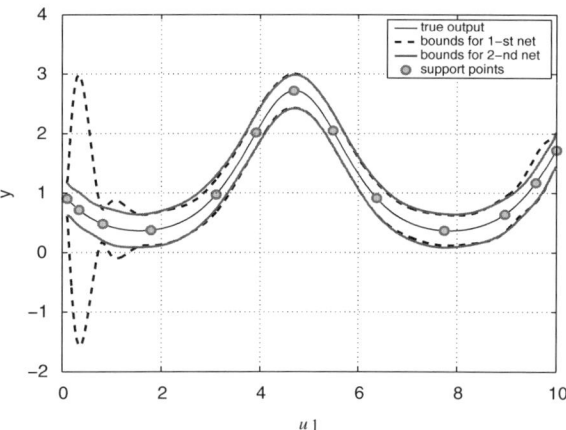

Fig. 3.6 Output and its bounds

Special Case

A special case that deserves particular attention is when the design consists of $n_e = n_p/m$ support points, i.e., the number of distinct support points equals that of parameters to be estimated. An example of such a design is presented in Sect. 3.1.2. The analysis of this example indicates that the weights associated with the support points are the same and equal $1/n_p$. This can be easily explained by transforming (3.41) into the following form:

$$P^{-1} = R^T W R, \tag{3.69}$$

where

$$R = \begin{bmatrix} R_1^T \\ \vdots \\ R_{n_e}^T \end{bmatrix} \tag{3.70}$$

and $W = \text{diag}(\mu_1 \mathbf{1}_m, \ldots, \mu_{n_e} \mathbf{1}_m)$.

It can be observed from (3.70) that R is a square matrix when $n_p = mn_e$. To achieve this, the number of support points should be

$$n_e = \frac{n_p}{m} = (n_h + 1) + \frac{n_h(n_r + 1)}{m}. \tag{3.71}$$

As can be seen from (3.71), the number of hidden neurons should be suitably selected to guarantee that n_e is a positive integer number. Thus, the determinant of (3.69) is

$$\det\left(P^{-1}\right) = \left(\prod_{k=1}^{n_e} \mu_k\right)^m \det(R)^2. \tag{3.72}$$

From (3.72), it is clear that μ_k, $k = 1, \ldots, n_p/m$, maximising $\det\left(P^{-1}\right)$ are the same and equal m/n_p.

Unfortunately, it is impossible to expect a priori how many support points should be used to form a D-optimum design for a given neural model. Indeed, the equation (3.71) indicates the minimum number of support points, while the maximum number can be determined with the help of Caratheodory's theorem [7, 10, 13] and is equal to $n_e = n_p(n_p + 1)/2 + 1$.

Now let us consider a single output neural model for which $n_e = n_p$. The parameter vector of (3.1) is estimated with the least-square method as follows:

$$\hat{p} = \arg\min_{p \in \mathbb{R}^{n_p}} \sum_{k=1}^{n_t} (y_k - h(p, u_k))^2. \tag{3.73}$$

Another appealing characteristic of the design considered can be expressed by the following theorem, which is based on the results presented in [23].

Theorem 3.4 *Assume that*

$$\xi = \left\{ \begin{matrix} u_1 & \cdots & u_{n_p} \\ \frac{1}{n_p} & \cdots & \frac{1}{n_p} \end{matrix} \right\},$$

and the number of observations for different u_ks is $n_x = \frac{n_L}{n_p}$ (it is assumed that it is a positive integer number). Assume also that for all $\beta \in \mathbb{R}^{n_p}$, there exists $p \in \mathbb{R}^{n_p}$ such that $\beta = [h(p, u_1), \ldots, h(p, u_{n_p})]^T$ and p does not satisfy the conditions of Theorem 3.1. Then the cost function of (3.73) has a unique global minimiser $\hat{p}$ and no other global minimisers.

Proof See the proof of Theorem 1 in [23].

Since the weights associated with support points are the same, then it is natural to assume that the number of observations for different u_ks is the same. Theorem 3.4 can be relatively easily interpreted because the optimisation problem (3.73) can be expressed (under the assumptions of Theorem 3.4) as

$$\min_{p \in \mathbb{R}^{n_p}} \sum_{k=1}^{n_p} (\bar{y}_k - h(p, u_k))^2 = 0, \tag{3.74}$$

and

$$\bar{y}_k = \frac{1}{n_x} \sum_{i=1}^{n_x} y_k^i, \tag{3.75}$$

where y_k^i stands for the ith observation under u_k. Thus, the solution $\hat{p}$ of (3.74) should satisfy

$$\bar{y}_k - h(\hat{p}, u_k) = 0, \quad k = 1, \ldots, n_p. \tag{3.76}$$

Indeed, the fact that $\hat{p}$ does not satisfy the conditions of Theorem 3.1 implies that (3.1) is uniquely determined by its input-output map, up to a finite group of symmetries (permutations of hidden neurons and changing the sign of all weights associated with a particular hidden neuron) [24]. This means that $\hat{p}$ is a unique solution of (3.76).

Towards Robustness: Sequential Design

As was shown in Sect. 3.1.2, the unappealing characteristic of experimental design for the MLP is the fact that the FIM depends on the non-linear parameters $P^{(n)}$ only. It is obvious that the true value of $P^{(n)}$ is unknown and hence its estimate should be utilised instead. As was mentioned in Sect. 3.1.2, if some rough estimates are given, i.e., they can be obtained with any training method for feed-forward neural networks [19], then the so-called *sequential design* [7, 13] can be applied. Such a strategy is usually applied off-line, i.e., the first step is parameter estimation while the second one is to use some specialised algorithms, e.g., the Wynn–Federov algorithm detailed in Sect. 3.1.2, to obtain a design of the form (3.26). In spite of the simplicity of such a sequential approach, some non-trivial problems arise which can be described as follows:

- Determination of the number of stages of experimentation-estimation required to attain the prescribed accuracy;
- Dependence of the final design upon the initial parameter estimates;
- Unique parametrisation of (3.1). This is the necessary condition to ensure the convergence of the sequential algorithm;
- Management of data collected in the consecutive experiments in order to guarantee the convergence of $\hat{p}$ to the true value of parameters p.

Some existing results being partial solutions to the first two questions can be found in [7, 25, 26]. Sussman [24] proved that, under some conditions, a network of the structure (3.1) with the hyperbolic tangent activation function is uniquely determined by its input-output map, up to a finite group of symmetries (permutations of hidden neurons and changing the sign of all weights associated with a particular hidden neuron). In [17] the author extended the results of [24] to the structure (3.1) with the logistic activation function. The solution to the last problem seems to be well developed and can be formulated as follows [7]: in order to guarantee the convergence of $\hat{p}$ to p, the estimation of $\hat{p}^k$ (the estimate of p in the kth iteration of the sequential algorithm) should make use of all previous observations collected during the preceding iterations of the sequential algorithm. Fukumizu [12] employed this strategy for OED for the MLP. The routine employed in [12] adds one single measurement to the measurement set collected in the preceding iterations of the algorithm. This new support point is obtained in such a way as to obtain an optimum design for the new parameter estimate. This idea is to be exploited in designing a new sequential algorithm that can be used for both training and data development for the MLP. Another approach [7] is to obtain a design for a new parameter estimate in a classic way, e.g., with the Wynn–Fedorov algorithm, while parameter estimation should make use of all the previous observations that where collected during the preceding iterations of the sequential algorithm. This strategy is employed in the numerical example presented in the subsequent section.

Numerical Example

Let us reconsider the example presented in Sect. 3.1.1. It is assumed that an initial parameter estimate is $\hat{p} = [1.8, 0.45, 0.54]^T$. The sequential algorithm utilises (3.30) to obtain OED in the consecutive iterations of the algorithm. The measurements y were generated by disturbing the data obtained with (3.28) by the normally distributed random noise $\mathcal{N}(0, 0.1^2)$. The Levenberg–Marquardt algorithm [7] was employed

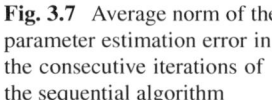

Fig. 3.7 Average norm of the parameter estimation error in the consecutive iterations of the sequential algorithm

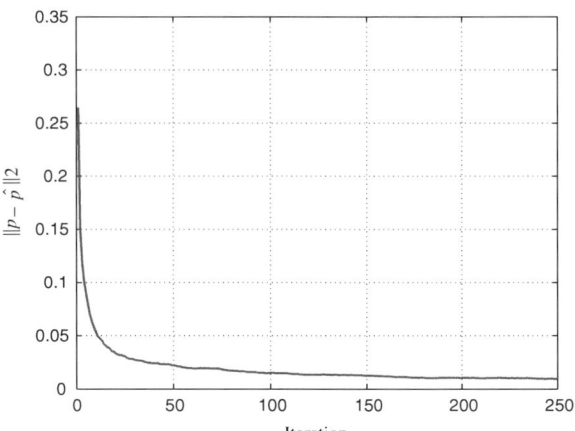

for parameter estimation. In order to show the reliability of the sequential algorithm, consisting of 250 cycles of estimation and experimentation, it was repeated 100 times. This means that in each cycle 9×250 measurements were collected, i.e., in each iteration the measurements were repeated three times for each support point of (3.30).

Figure 3.7 shows an average norm of the parameter estimation error $\|\boldsymbol{p} - \hat{\boldsymbol{p}}\|_2$ in the consecutive iterations of the sequential algorithm. From this result it can be seen that the parameter estimate converges (on average) to the true parameter vector $\boldsymbol{p}$. This implies that the designed experiment tends to the optimal experiment for $\boldsymbol{p}$.

3.1.3 Industrial Application

This section presents an industrial application study regarding the proposed approach. In particular, the presented example concerns experimental design, neural model development and fault detection of a valve actuator. The problem regarding FDI of this actuator was attacked from different angles in the EU DAMADICS project. *Development and Application of Methods for Actuator Diagnosis in Industrial Control Systems* (DAMADICS) was a research project focused on drawing together wide-ranging techniques and fault diagnosis within the framework of real application to on-line diagnosis of a 5-stage evaporisation plant of the sugar factory in Lublin, Poland. The project was focused on the diagnosis of valve (cf. Fig. 3.8) plant actuators and looked towards real implementation methods for new actuator systems. The sugar factory was a subcontractor (under the Warsaw University of Technology) providing real process data and the evaluation of trials of fault diagnosis methods.

The control valve is a mean used to prevent, permit and/or limit the flow of sugar juice through the control system (a detailed description of this actuator can be found

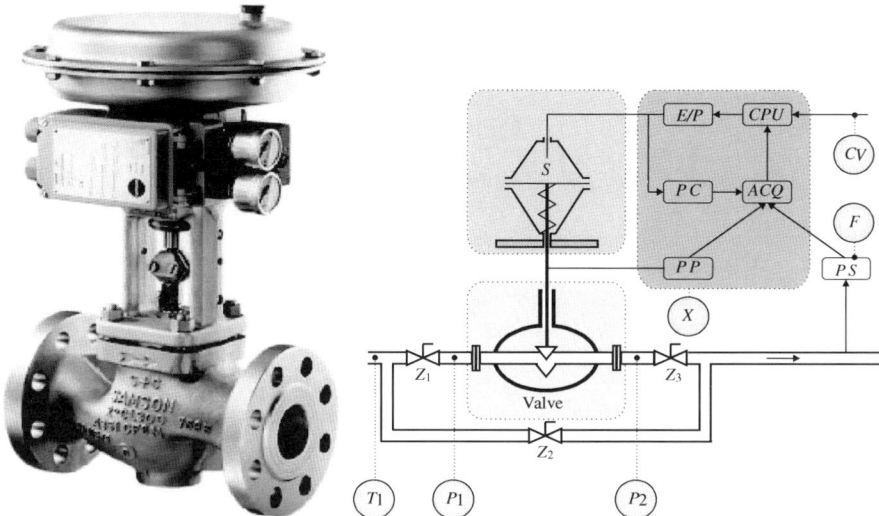

Fig. 3.8 Actuator and its scheme

in [27]). As can be seen in Fig. 3.8, the following process variables can be measured: CV is the control signal, $P1$ is the pressure at the inlet of the valve, $P2$ is the pressure at the outlet of the valve, $T1$ is the juice temperature at the inlet of the valve, X is servomotor rod displacement, F is the juice flow at the outlet of the valve. Thus, the output is $y = (F, X)$, while the input is given by $u = (CV, P1, P2, T)$. In Fig. 3.8, three additional bypass valves (denoted by z_1, z_2, and z_3) can be seen. The state of these valves can be controlled manually by the operator. They are introduced for manual process operation, actuator maintenance and safety purposes. The data gathered from the real plant can be found on the DAMADICS website [27]. Although a large amount of real data is available, they do not cover all faulty situations, while the simulator is able to generate a set of 19 faults (see Table 3.1) Moreover, due to a strict production regime, operators do not allow changing plant inputs, i.e., they are set up by control systems. Thus, an actuator simulator was developed with MATLAB Simulink (available at [27]). Apart from experimental design purposes, the main reason for using the data from the simulator is the fact that the achieved results can be easily compared with those obtained with different approaches, e.g., [28–30]. The main objective of the subsequent part of this section is to develop a neural network that can be used for fault detection of the industrial valve actuator. The above task can be divided into the following steps:

Step 1: Training of a network based on the nominal data set;
Step 2: Design of the experiment with the Wynn–Fedorov algorithm described in Sect. 3.1.2 based on the network obtained in *Step 1*;
Step 3: Training of a network based on the data obtained with experimental design.

Table 3.1 Set of faults considered for the benchmark

Fault	Description	S	M	B	I
f_1	Valve clogging	x	x	x	
f_2	Valve plug or valve seat sedimentation			x	x
f_3	Valve plug or valve seat erosion				x
f_4	Increased valve or busing friction				x
f_5	External leakage				x
f_6	Internal leakage (valve tightness)				x
f_7	Medium evaporation or critical flow	x	x	x	x
f_8	Twisted servomotor's piston rod	x	x	x	
f_9	Servomotor housing or terminal tightness				x
f_{10}	Servomotor's diaphragm perforation	x	x	x	
f_{11}	Servomotor's spring fault				x
f_{12}	Electro-pneumatic transducer fault	x	x	x	
f_{13}	Rod displacement sensor fault	x	x	x	x
f_{14}	Pressure sensor fault	x	x	x	
f_{15}	Positioner feedback fault			x	
f_{16}	Positioner supply pressure drop	x	x	x	
f_{17}	Unexpected pressure change across the valve			x	x
f_{18}	Fully or partly opened bypass valves	x	x	x	x
f_{19}	Flow rate sensor fault	x	x	x	

Abrupt faults: S small, M medium, B big, I incipient faults

Based on the experience with the industrial valve actuator, it was observed that the following subset of the measured variables is sufficient for fault detection purposes: $u = (CV, P1, 1)$, $y = F$.

In Step 1, a number of experiments (the training of a neural network with the Levenberg–Marquardt algorithm [7]) were performed in order to find a suitable number of hidden neurons n_h (cf. (3.1)). For that purpose, $n_t = 100$ data points were generated for which inputs were uniformly spread within the design region $\mathbb{U}$, where $0.25 < u_1 < 0.75$ and $0.6625 < u_2 < 0.8375$. As a result, a neural model consisting of $n_h = 5$ hidden neurons was obtained. The main objective of Step 2 was to utilise the above model and the Wynn–Fedorov algorithm in order to obtain D-optimum experimental conditions. First, an initial experiment was generated in such a way as to equally divide the design space of u. Finally, the Wynn–Fedorov algorithm was applied. Figure 3.9 shows the support points ($n_e = 45$) and the variance function for the obtained D-optimum design. Based on the derived continuous design, a set consisting of $n_t = 100$ points was found and used for data generation. The number of repetitions of each optimal input u_k was calculated by suitably rounding the numbers $\mu_k n_t$, $k = 1, \ldots, n_e$ [7, 10]. It should be strongly stressed that the data were collected in the steady-state of the valve because the utilised model (3.1) was static. Finally, the new data set was used for training the network with the Levenberg–Marquardt algorithm. As was mentioned at the beginning of this section, the research directions of the DAMADICS project were oriented towards fault diagnosis and, in particular, fault detection of the valve actuator. Under the assumption of a perfect

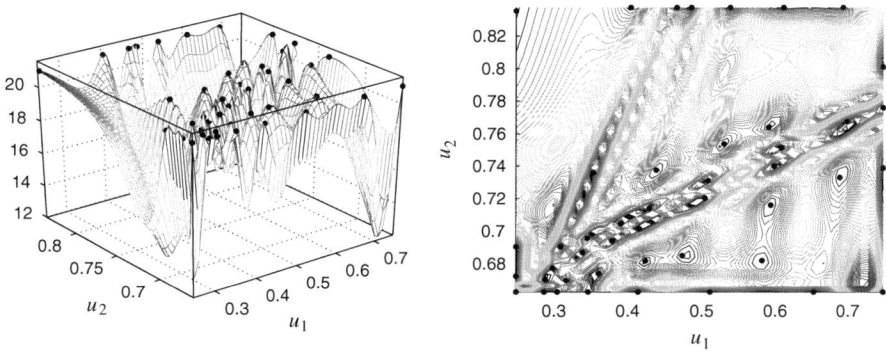

Fig. 3.9 Variance function and the corresponding support points

mathematical description of the systems considered, a perfect residual generation should provide a residual that is zero during the normal operation of the system and considerably different than zero otherwise. This means that the residual should ideally carry information regarding a fault only. Under such assumptions, faults can be easily detected. Unfortunately, this is impossible to attain in practice since residuals are normally uncertain, corrupted by noise, disturbances and modelling uncertainty. That is why, in order to avoid false alarms, it is necessary to assign a threshold to the residual that is significantly larger than zero. The most common approach is to use a fixed threshold [22, 31]. The main difficulty with this kind of thresholds is the fact that they may cause many serious problems regarding false alarms as well as undetected faults. In other words, it is very difficult to fix such a threshold and there is no optimal solution that can be applied to settle such a task. Fortunately, using (3.17), (3.24) and (3.25), it is possible to develop an adaptive threshold that can be described as follows:

$$|z_{i,k}| \leq t_{n_t-n_p}^{\alpha/2} \hat{\sigma} \left(1 + \boldsymbol{r}_{i,k} \boldsymbol{P} \boldsymbol{r}_{i,k}^T\right)^{1/2}, \quad i = 1, \ldots, m. \tag{3.77}$$

Consequently, the decision logic can be realised as follows:

If the residual z_k satisfies (3.77), *then there is no fault symptom, else* (3.77) *indicates that a fault symptom has occurred.*

The objective of the subsequent part of this section is to use the obtained network for fault detection, as well as to compare its performance with that of a network obtained for a nominal data set. Table 3.2 shows the results of fault detection for a set of faults being specified for the benchmark (the symbols S—Small, M—Medium, and B—Big denote the magnitude of the faults). All faults were generated with the same scenario, i.e., the first 200 samples correspond to the normal operating mode of the system while the remaining ones were generated under faulty conditions. Figure 3.10 presents the residual signal obtained with a network trained with the D-optimum data set as well as an adaptive threshold provided by this network (the 2nd network). This figure also presents an adaptive threshold provided by a network (the 1st network)

Table 3.2 Results of fault detection

Fault	Description	S	M	B
f_1	Valve clogging	D	D	D
f_2	Valve plug or valve seat sedimentation	X	X	D
f_7	Medium evaporation or critical flow	D	D	D
f_8	Twisted servomotor's piston rod	N	N	N
f_{10}	Servomotor's diaphragm perforation	D	D	D
f_{12}	Electro-pneumatic transducer fault	X	X	D
f_{13}	Rod displacement sensor fault	D	D	D
f_{15}	Positioner feedback fault	X	X	D
f_{16}	Positioner supply pressure drop	N	N	D
f_{17}	Unexpected pressure change across valve	X	X	D
f_{18}	Fully or partly opened bypass valves	D	D	D
f_{19}	Flow rate sensor fault	D	D	D

D detected, N not detected, X not specified for the benchmark

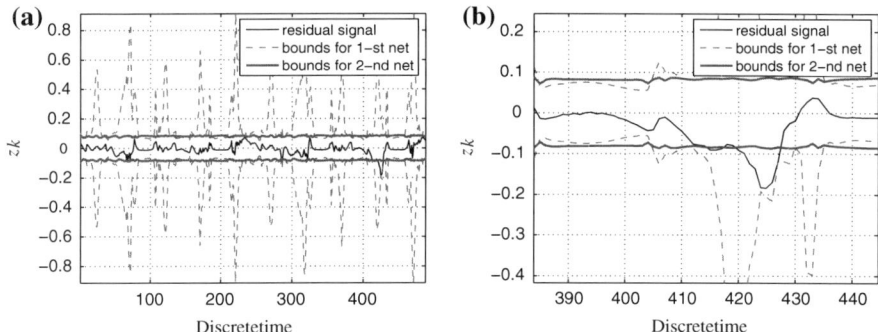

Fig. 3.10 Residual and adaptive thresholds for the fault f_1—small (**a**) and its selected part (**b**)

trained with the data set generated by equally dividing the design space. It can be observed that the neural network obtained with the use of D-optimum experimental design makes it possible to obtain more accurate bounds than those obtained with a neural network trained otherwise. Indeed, as can be seen in Fig. 3.10b, the fault f_1—small (which in the light of its nature is hard to detect) can be detected with the help of the 2nd network while it is impossible to detect it with the use of the 1st one. It should be strongly underlined that the situation is even worse when the 1st network is used for residual generation, i.e., in the presented example it was used for adaptive threshold generation only. As can be observed in Table 3.2, almost all the faults specified for the benchmark can be detected. The main reason why the faults f_8 and f_{16} (small and medium) cannot be detected is because their effect is exactly at the same level as that of noise. However, it should be pointed out that this was the case for other techniques [29, 30] tested with the DAMADICS benchmark. Finally, Fig. 3.11 presents sample residuals for the faults f_{18}—small and f_{19}—small, respectively.

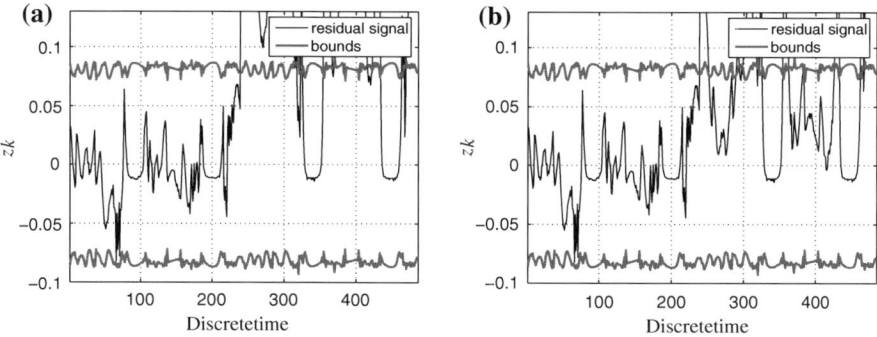

Fig. 3.11 Residual and adaptive thresholds for the fault f_{18}—small (**a**) and f_{19}—small (**b**), respectively

3.2 GMDH Neural Networks

The synthesis process of the GMDH neural network [8] is based on the iterative processing of a sequence of operations. This process leads to the evolution of the resulting model structure in such a way as to obtain the best quality approximation of the real system. The quality of the model can be measured with the application of various criteria [32]. The resulting GMDH neural network is constructed through the connection of a given number of neurons, as shown in Fig. 3.12. The neuron has the following structure:

$$y_{n,k}^{(l)} = \xi\left(\left(r_{n,k}^{(l)}\right)^T p_n^{(l)}\right), \qquad (3.78)$$

where $y_{n,k}^{(l)}$ stands for the neuron output (l is the layer number, n is the neuron number in the lth layer), whilst $\xi(\cdot)$ denotes a non-linear invertible activation function, i.e., there exists $\xi^{-1}(\cdot)$. Moreover, $r_{n,k}^{(l)} = g\left(\left[u_{i,k}^{(l)}, u_{j,k}^{(l)}\right]^T\right)$, $i, j = 1, \ldots, r$, and $p_n^{(l)} \in \mathbb{R}^{n_p}$ are the regressor and the parameter vectors, respectively, and $g(\cdot)$ is an arbitrary bivariate vector function, e.g., $g(x) = [x_1^2, x_2^2, x_1 x_2, x_1, x_2, 1]^T$ that corresponding to the bivariate polynomial of the second degree.

An outline of the GMDH algorithm can be as follows [8, 33]:

Step 1: Determine all neurons (estimate their parameter vectors $p_n^{(l)}$ with the training data set $\mathcal{T}$) whose inputs consist of all possible couples of input variables, i.e., $(r-1)r/2$ couples (neurons).

Step 2: Using a validation data set $\mathcal{V}$, not employed during the parameter estimation phase, select several neurons which are best fitted in terms of the chosen criterion.

Step 3: If the termination condition is fulfilled (either the network fits the data with desired accuracy, or the introduction of new neurons did not induce a significant increase in the approximation abilities of the neural network), then STOP, otherwise

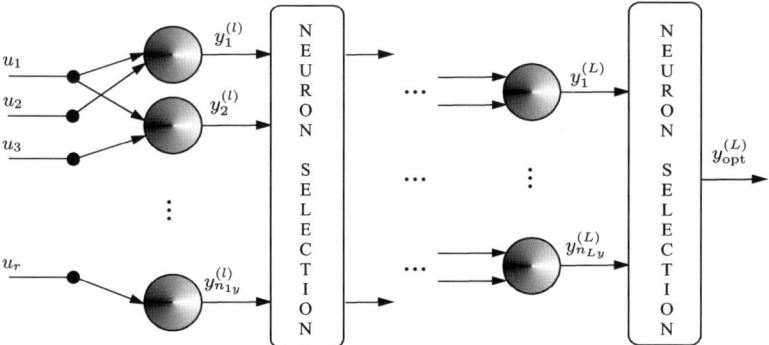

Fig. 3.12 Principle of the GMDH algorithm

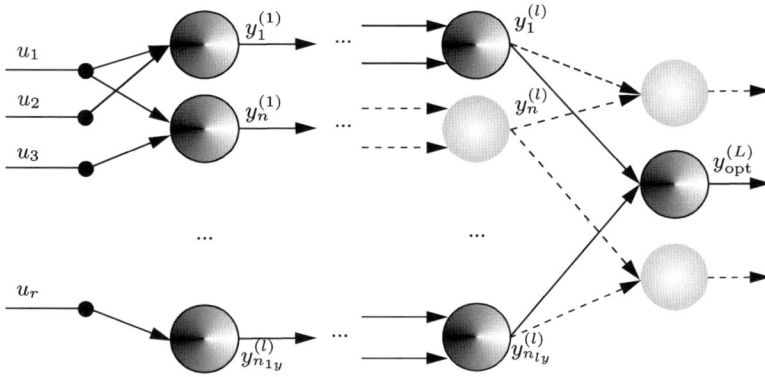

Fig. 3.13 Final structure of the GMDH neural network

use the outputs of the best-fitted neurons (selected in *Step 2*) to form the input vector for the next layer (see Fig. 3.12), and then go to Step 1.

To obtain the final structure of the network (Fig. 3.13), all unnecessary neurons are removed, leaving only those which are relevant to the computation of the model output. The procedure of removing the unnecessary neurons is the last stage of the synthesis of the GMDH neural network. The feature of the above algorithm is that the techniques for parameter estimation of linear-in-parameter models can be used during the realisation of Step 1. Indeed, since $\xi(\cdot)$ is invertible, the neuron (3.78) can relatively easily be transformed into a linear-in-parameter one.

3.2.1 Model Uncertainty in the GMDH Neural Network

The objective of system identification is to obtain a mathematical description of a real system based on input-output measurements. Irrespective of the identification method used, there is always the problem of model uncertainty, i.e., the model-reality mismatch. Even though the application of the GMDH approach to model structure selection can improve the quality of the model, the resulting structure is not the same as that of the system. It can be shown [34] that the application of the classic evaluation criteria like the Akaike Information Criterion (AIC) and the Final Prediction Error (FPE) [32, 35] can lead to the selection of inappropriate neurons and, consequently, to unnecessary structural errors.

Apart from the model structure selection stage, inaccuracy in parameter estimates also contributes to modelling uncertainty. Indeed, while applying the least-square method to parameter estimation of neurons (3.78), a set of restrictive assumptions has to be satisfied. The first (and the most controversial) assumption is that the structure of the neuron is the same as that of the system (no structural errors). In the case of the GMDH neural network, this condition is extremely difficult to satisfy. Indeed, neurons are created based on two input variables selected from U and hence it is impossible to eliminate the structural error. Another assumption concerns the transformation with $\xi^{-1}(\cdot)$. Let us consider the following system output signal:

$$y_k = \xi\left(\left(r_{n,k}^{(l)}\right)^T p_n^{(l)}\right) + v_{n,k}^{(l)}. \tag{3.79}$$

The use of linear-in-parameter estimation methods for the model (3.78), e.g., the Least-Square Method (LSM) [36] requires transforming the output of the system (3.79) as follows:

$$\left(r_{n,k}^{(l)}\right)^T p_n^{(l)} = \xi^{-1}(y_k) - v_{n,k}^{(l)}. \tag{3.80}$$

Unfortunately, the transformation of (3.79) with $\xi^{-1}(\cdot)$ results in

$$\left(r_{n,k}^{(l)}\right)^T p_n^{(l)} = \xi^{-1}\left(y_k - v_{n,k}^{(l)}\right). \tag{3.81}$$

Thus, good results can only be expected when the noise $v_{n,k}^{(l)}$ magnitude is relatively small. The other assumptions are directly connected with the properties of the LSM, i.e., in order to attain an estimator $\hat{p}_{n,k}^{(l)}$ of $p_{n,k}^{(l)}$ for (3.78) which is unbiased and minimum variance [10], it is assumed that

$$\mathcal{E}\left[v_n^{(l)}\right] = \mathbf{0}, \tag{3.82}$$

$$\mathrm{cov}\left[v_n^{(l)}\right] = \left(\sigma_n^{(l)}\right)^2 \mathbf{I}. \tag{3.83}$$

The assumption (3.82) means that there are no structural errors (deterministic disturbances) and model uncertainty is described in a purely stochastic way (uncorrelated noise, cf. (3.83)). It must be pointed out that this is rather difficult to satisfy this condition in practice.

Let us suppose that, in some case, the conditions (3.82) and (3.83) are satisfied. Then it can be shown that $\hat{p}_{n,k}^{(1)}$ (the parameter estimate vector for a neuron of the first layer) is unbiased and minimum variance [10]. Consequently, the neuron output in the first layer becomes the input to other neurons in the second layer. The system output estimate can be described by

$$\hat{y}_n^{(l)} = R_n^{(l)} \left[\left(R_n^{(l)} \right)^T R_n^{(l)} \right]^{-1} \left(R_n^{(l)} \right)^T Y, \tag{3.84}$$

where $R_n^{(l)} = [r_{n,1}^{(l)}, \ldots, r_{n,n_t}^{(l)}]^T$, $Y = [y_1, \ldots, y_{n_t}]^T$, and $\hat{y}_n^{(l)} = [\hat{y}_{n,1}^{(l)}, \ldots, \hat{y}_{n,n_t}^{(l)}]^T$ represent the system output vector and its estimate. Apart from the situation in the first layer ($l = 1$), where the matrix $R_n^{(l)}$ depends on U, in the subsequent layers $R_n^{(l+1)}$ depends on (3.84) and hence

$$\mathcal{E}\left[\left[\left(R_n^{(l+1)} \right)^T R_n^{(l+1)} \right]^{-1} \left(R_n^{(l+1)} \right)^T v_n^{(l+1)} \right] \neq 0. \tag{3.85}$$

That is why the parameter estimator in the next layers is biased and no minimum variance, i.e.,

$$\begin{aligned}
\mathcal{E}\left[\hat{p}_n^{(l+1)} \right] &= \mathcal{E}\left[\left[\left(R_n^{(l+1)} \right)^T R_n^{(l+1)} \right]^{-1} \left(R_n^{(l+1)} \right)^T y \right] \\
&= \mathcal{E}\left[\left[\left(R_n^{(l+1)} \right)^T R_n^{(l+1)} \right]^{-1} \left(R_n^{(l+1)} \right)^T \left(R_n^{(l+1)} p_n^{(l+1)} + v_n^{(l+1)} \right) \right] \\
&= p_n^{(l+1)} + \mathcal{E}\left[\left[\left(R_n^{(l+1)} \right)^T R_n^{(l+1)} \right]^{-1} \left(R_n^{(l+1)} \right)^T v_n^{(l+1)} \right]. \tag{3.86}
\end{aligned}$$

To settle this problem, the instrumental variable method or other methods listed in [7] can be employed. On the other hand, these methods provide only asymptotic convergence, and hence a large data set is usually required to obtain an unbiased parameter estimate.

3.2.2 Bounded-Error Approach

The problems detailed in the previous section clearly show that there is a need for the application of a parameter estimation method different than the LSM. Such a method should also be easily adaptable to the case of an uncertain regressor and should overcome all of the remaining difficulties mentioned in Sect. 3.2.1. The subsequent part of this section gives an outline of such a method called the Bounded-Error Approach (BEA).

Bounded Noise/Disturbances

The usual statistical parameter estimation framework assumes that data are corrupted by errors which can be modelled as realisations of independent random variables, with a known or parameterised distribution. A more realistic approach is to assume that the errors lie between given prior bounds. This is the case, for example, for data collected with an analogue-to-digital converter or for measurements performed with a sensor of a given type. Such reasoning leads directly to the bounded-error approach [6, 7]. Let us consider the following system:

$$y_k = \left(r_{n,k}^{(l)}\right)^T p_n^{(l)} + v_{n,k}^{(l)}. \tag{3.87}$$

The problem is to obtain the parameter estimate $\hat{p}_n^{(l)}$ as well as an associated parameter uncertainty required to design a robust fault detection system. In order to simplify the notation, the index $_n^{(l)}$ is omitted. The knowledge regarding the set of admissible parameter values allows obtaining the confidence interval of the model output which satisfies

$$y_k^N \leq y_k \leq y_k^M, \tag{3.88}$$

where y_k^N and y_k^M are respectively the minimum and maximum admissible values of the model output that are consistent with the input-output measurements of the system. Under the assumptions detailed in Sect. 3.2.1, the uncertainty of the neural network can be obtained according to [37].

In this work, it is assumed that v_k is bounded as follows:

$$v_k^N \leq v_k \leq v_k^M, \tag{3.89}$$

while the bounds v_k^N and v_k^M ($v_k^N \neq v_k^M$) are known a priori. The idea underlying the bounded-error approach is to obtain a feasible parameter set [6]. It can be defined as

$$\mathbb{P} = \left\{p \in \mathbb{R}^{n_p} \mid y_k - v_k^M \leq r_k^T p \leq y_k - v_k^N, k = 1, \ldots, n_t\right\}. \tag{3.90}$$

This set can be perceived as a region of the parameter space that is determined by n_t pairs of hyperplanes:

$$\mathbb{P} = \bigcap_{k}^{n_t} \mathbb{S}_k, \tag{3.91}$$

where each pair defines the parameter strip

Fig. 3.14 Feasible parameter set

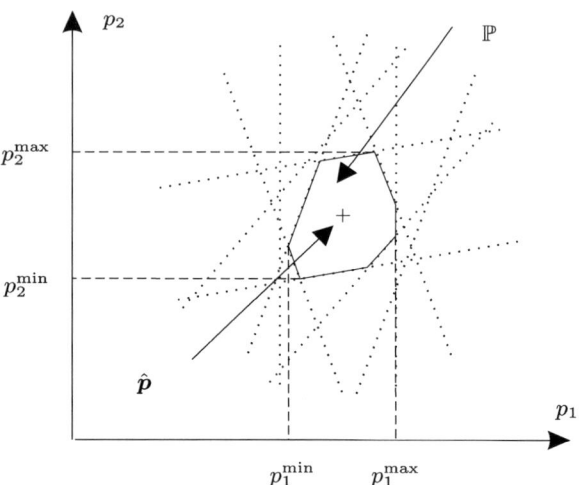

$$\mathbb{S}_k = \left\{ \boldsymbol{p} \in \mathbb{R}^{n_p} \mid y_k - v_k^M \le \boldsymbol{r}_k^T \boldsymbol{p} \le y_k - v_k^N \right\}. \tag{3.92}$$

Any parameter vector contained in $\mathbb{P}$ is a valid estimate of $\boldsymbol{p}$. In practice, the centre (in some geometrical sense) of $\mathbb{P}$ (cf. Fig. 3.14 for $n_p = 2$) is chosen as the parameter estimate $\hat{\boldsymbol{p}}$, e.g.,

$$\hat{p}_i = \frac{p_i^{\min} + p_i^{\max}}{2}, \quad i = 1, \dots, n_p, \tag{3.93}$$

with

$$p_i^{\min} = \arg\min_{p \in \mathbb{P}} p_i, \quad i = 1, \dots, n_p, \tag{3.94}$$

$$p_i^{\max} = \arg\max_{p \in \mathbb{P}} p_i, \quad i = 1, \dots, n_p. \tag{3.95}$$

This is, of course, important when the task is to develop a neural network for which the knowledge regarding parameter uncertainty is not useful. In the presented approach, the nominal model is obtained in such a way that the knowledge regarding parameter uncertainty is used for fault detection purposes.

The problems (3.94) and (3.95) can be solved with the well-known linear programming techniques [6, 38], but when n_t and/or n_p are large, the computational cost may be significant. This constitutes the main drawback of the approach. One way out of this problem is to apply a technique where constraints are executed separately one after another [39], although this approach does not constitute a perfect remedy for the computational problem considered. This means that the described BEA can be employed for tasks with a relatively small dimension, as is the case for GMDH neurons. In spite of the above-mentioned computational problems, the technique described in [39] was implemented and used in this work. The main difficulty associated with the BEA concerns a priori knowledge regarding the error bounds v_k^N and v_k^M. However, these bounds can also be estimated [6] by assuming that $v_k^N = v^N$, $v_k^M = v^M$, and then suitably extending the unknown parameter vector p by v^N and v^M. Determining the bounds can now be formulated as

$$(v^N, v^M) = \arg \min_{v^M \geq 0, \ v^N \leq 0} v^M - v^N, \qquad (3.96)$$

with respect to the following constraints:

$$y_k - v^M \leq r_k^T p \leq y_k - v^N, \quad k = 1, \ldots, n_t. \qquad (3.97)$$

In this section, the well-known simplex method was utilised to solve the problem (3.96). Then, knowing v^N and v^M, the strategy described in [39] was employed.

Model Output Uncertainty

The methodology described in Sect. 3.2.2 makes it possible to obtain the parameter estimate $\hat{p}$ and the associated feasible parameter set $\mathbb{P}$. However, from a practical point of view, it is more convenient to obtain the system output confidence interval, i.e., the interval in which the "true" model output $y(k)$ can be found. This kind of knowledge makes it possible to obtain an adaptive threshold [40], and hence to develop a fault diagnosis scheme that is robust to model uncertainty.

Let $\mathbb{V}$ be the set of all vertices p^i, $i = 1, \ldots, n_v$, describing the feasible parameter set $\mathbb{P}$ (cf. (3.91)). If there is no error in the regressor, then the problem of determining the model output confidence interval can be solved as follows:

$$y_{M,k}^N = r_k^T p_k^N \leq r_k^T p \leq r_k^T p_k^M = y_{M,k}^M, \qquad (3.98)$$

where

$$p_k^N = \arg \min_{p \in \mathbb{V}} r_k^T p, \qquad (3.99)$$

$$p_k^M = \arg \max_{p \in \mathbb{V}} r_k^T p. \qquad (3.100)$$

The computation of (3.99) and (3.100) is realised by multiplying the parameter vectors corresponding to all vertices belonging to $\mathbb{V}$ by r_k^T.

Since (3.98) describes a neuron output confidence interval, the system output will satisfy

$$r_k^T p_k^N + v_k^N \leq y_k \leq r_k^T p_k^M + v_k^M. \qquad (3.101)$$

A more general case of (3.101) for neurons with a non-linear activation function will be considered in Sect. 3.2.3. The neuron output confidence interval defined by (3.98) and the corresponding system output confidence interval (3.101) are presented in

Fig. 3.15 Model output
confidence interval for the
error-free regressor

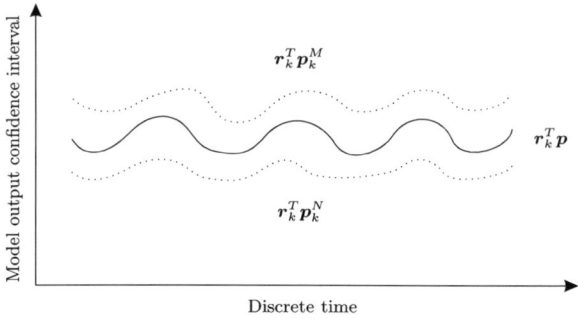

Fig. 3.16 System output
confidence interval for the
error-free regressor

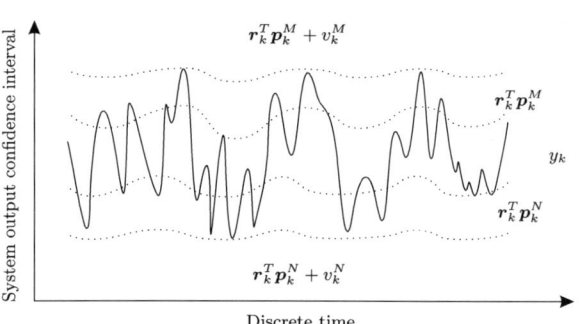

Figs. 3.15 and 3.16, respectively. As has already been mentioned, the neurons in the
lth ($l > 1$) layer are fed with the outputs of the neurons from the ($l − 1$)th layer.
Since (3.98) describes the model output confidence interval, the parameters of the
neurons in the layers have to be obtained with an approach that solves the problem
of an uncertain regressor [6].

   In order to modify the approach presented in Sect. 3.2.2, let us denote an unknown
"true" value of the regressor $\boldsymbol{r}_{n,k}$ by a difference between a known (measured) value
of the regressor $\boldsymbol{r}_k$ and the error in the regressor $\boldsymbol{e}_k$:

$$\boldsymbol{r}_{n,k} = \boldsymbol{r}_k - \boldsymbol{e}_k, \tag{3.102}$$

where it is assumed that the error $\boldsymbol{e}_k$ is bounded as follows:

$$e_{i,k}^N \leq e_{i,k} \leq e_{i,k}^M, \quad i = 1, \ldots, n_p. \tag{3.103}$$

Using (3.87) and substituting (3.102) into (3.103), one can define the region containing parameter estimates:

$$v_k^N - \boldsymbol{e}_k^T \boldsymbol{p} \leq y_k - \boldsymbol{r}_k^T \boldsymbol{p} \leq v_k^M - \boldsymbol{e}_k^T \boldsymbol{p}. \tag{3.104}$$

Unfortunately, for the purpose of parameter estimation it is not enough to introduce (3.102) into (3.103). Indeed, the bounds of (3.104) depend also on the sign of each p_i. It is possible to directly obtain these signs only for models whose parameters have physical meaning [41]. For models such as GMDH neural networks this is rather impossible. In [6, Chapters 17 and 18], the authors proposed some heuristic techniques, but these drastically complicate the problem (3.104) and do not seem to guarantee that these signs will be obtained properly. Bearing in mind the fact that the neuron (3.78) contains only a few parameters, it is possible to replace them by

$$p_i = p_i' - p_i'', \quad p_i', p_i'' \geq 0, \quad i = 1, \ldots, n_p. \tag{3.105}$$

Although the above solution is very simple, it doubles the number of parameters, i.e., instead of estimating n_p parameters it is necessary to do so for $2n_p$ parameters. In spite of that, this technique is very popular and widely used in the literature [6, 42]. Due to the above solution, (3.104) can be modified as follows:

$$\begin{aligned} v_k^N - \left(e_k^M\right)^T p' + \left(e_k^N\right)^T p'' \\ \leq y_k - r_k^T(p' - p'') \\ \leq v_k^M - \left(e_k^N\right)^T p' + \left(e_k^M\right)^T p''. \end{aligned} \tag{3.106}$$

This transformation makes it possible to employ, with a minor modification, the approach described in Sect. 3.2.2. The difference is that the algorithm processes each constraint (associated with a pair of hyperplanes defined with (3.106)) separately. The reason for such a modification is that the hyperplanes are not parallel [43].

The proposed modification of the BEA makes it possible to estimate the parameter vectors of the neurons from the lth, $l > 1$, layer. In the case of an error in the regressor, using (3.106) it can be shown that the model output confidence interval has the following form:

$$y_{M,k}^N \left(p_k^{'N}, p_k^{''N}\right) \leq r_n^T p \leq y_{M,k}^M \left(p_k^{'M}, p_k^{''M}\right), \tag{3.107}$$

where

$$y_{M,k}^N \left(p_k^{'N}, p_k^{''N}\right) = \left(r_k - e_k^M\right)^T p_k^{'N} + \left(e_k^N - r_k\right)^T p_k^{''N}, \tag{3.108}$$

$$y_{M,k}^M \left(p_k^{'M}, p_k^{''M}\right) = \left(r_k - e_k^N\right)^T p_k^{'M} + \left(e_k^M - r_k\right)^T p_k^{''M}, \tag{3.109}$$

and

$$\left(p_k^{'N}, p_k^{''N}\right) = \arg \min_{(p_k', p_k'') \in \mathbb{V}} y_{M,k}^N (p_k', p_k''), \tag{3.110}$$

$$\left(\boldsymbol{p}_k^{\prime M}, \boldsymbol{p}_k^{\prime\prime M} \right) = \arg \max_{(\boldsymbol{p}_k', \boldsymbol{p}_k'') \in \mathbb{V}} y_{M,k}^M (\boldsymbol{p}_k', \boldsymbol{p}_k''). \tag{3.111}$$

Using (3.107), it is possible to obtain the system output confidence interval:

$$y_{M,k}^N \left(\boldsymbol{p}_k^{\prime N}, \boldsymbol{p}_k^{\prime\prime N} \right) + v_k^N \leq y_k \leq y_{M,k}^M \left(\boldsymbol{p}_k^{\prime M}, \boldsymbol{p}_k^{\prime\prime M} \right) + v_k^M. \tag{3.112}$$

3.2.3 Synthesis of the GMDH Neural Network Via the BEA

In order to adapt the approach of Sect. 3.2.2 to parameter estimation of (3.78), it is necessary to transform the relation

$$v_k^N \leq y_k - \xi \left(\left(\boldsymbol{R}_{n,k}^{(l)} \right)^T \boldsymbol{p}_n^{(l)} \right) \leq v_k^M \tag{3.113}$$

in such a way as to avoid the problems detailed in Sect. 3.2.1. In this case, it is necessary to assume that

1. $\xi(\cdot)$ is continuous and bounded, i.e.,

$$\forall x \in \mathbb{R} : a < \xi(x) < b; \tag{3.114}$$

2. $\xi(\cdot)$ is monotonically increasing, i.e.,

$$\forall x, y \in \mathbb{R} : x \leq y \text{ iff } \xi(x) \leq \xi(y). \tag{3.115}$$

Now it is easy to show that

$$y_k - v_k^M \leq \xi \left(\left(\boldsymbol{R}_{n,k}^{(l)} \right)^T \boldsymbol{p}_n^{(l)} \right) \leq y_k - v_k^N, \tag{3.116}$$

and then

$$\xi^{-1} \left(y_k - v_k^M \right) \leq \left(\boldsymbol{R}_{n,k}^{(l)} \right)^T \boldsymbol{p}_n^{(l)} \leq \xi^{-1} \left(y_k - v_k^N \right). \tag{3.117}$$

As was pointed out in Sect. 3.2.2, an error in the regressor must be taken into account during the design procedure of the neurons from the second and subsequent layers. Indeed, by using (3.98) in the first layer and (3.107) in the subsequent ones it is possible to obtain the bounds of the output (3.78) and the bounds of the regressor error (3.89), whilst the known value of the regressor should be computed by using the parameter estimates $\hat{\boldsymbol{p}}_n^{(l)}$. Note that the processing errors of the neurons, which are described by the model output confidence interval (3.107), can be propagated and accumulated during the introduction of new layers (Fig. 3.17). This unfavourable phenomenon can be reduced by the application of an appropriate selection method [44].

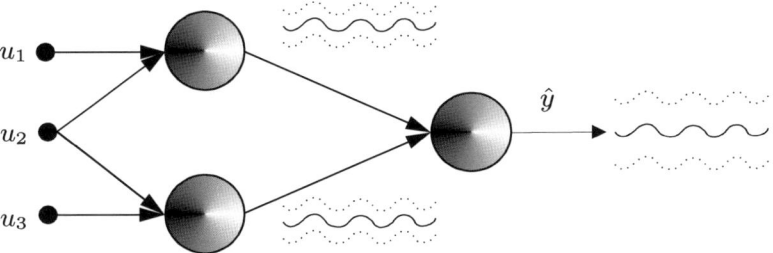

Fig. 3.17 Propagation of model uncertainty (*dotted lines*), model response (*continuous line*)

Selection methods in GMDH neural networks play the role of a mechanism of structural optimisation at the stage of constructing a new layer of neurons. Only well-performing neurons are preserved to build a new layer. During the selection, neurons which have too large a value of the defined evaluation criteria [32, 35, 44] are rejected based on chosen selection methods. Unfortunately, as was mentioned in Sect. 3.2.1, the application of the classic evaluation criteria like the Akaike Information Criterion (AIC) and the Final Prediction Error (FPE) [32, 35] during network synthesis may lead to the selection of an inappropriate structure of the GMDH neural network. This follows from the fact that the above criteria do not take into account modelling uncertainty. In this way, neurons with small values of the classic quality indexes Q_V but with large uncertainty (Fig. 3.18) can be obtained. In order to overcome this difficulty, a new evaluation criterion of the neurons has been introduced in [8], i.e.,

$$Q_V = \frac{1}{n_V} \sum_{k=1}^{n_v} \left| \left(y_{M,k}^M + v_k^M \right) - \left(y_{M,k}^N + v_k^N \right) \right|, \tag{3.118}$$

where n_V is the number of input-output measurements for the validation data set, while y_k^M and y_k^N are calculated with (3.98) for the first layer or with (3.108)–(3.109) for the subsequent ones. Finally, the neuron in the last layer that gives the smallest processing error (3.118) constitutes the output of the GMDH neural network, while model uncertainty of this neuron is used for the calculation of the system output confidence interval. It is therefore possible to design the so-called adaptive threshold [40], which can be employed for robust fault detection.

3.2.4 Robust Fault Detection with the GMDH Model

The purpose of this section is to show how to develop an adaptive threshold with the GMDH model and some knowledge regarding its uncertainty. Since the residual is

$$z_k = y_k - \hat{y}_k, \tag{3.119}$$

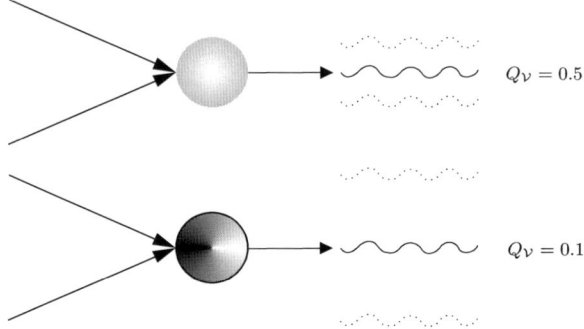

$Q_V = 0.5$

$Q_V = 0.1$

Fig. 3.18 Problem of an incorrect selection of a neuron

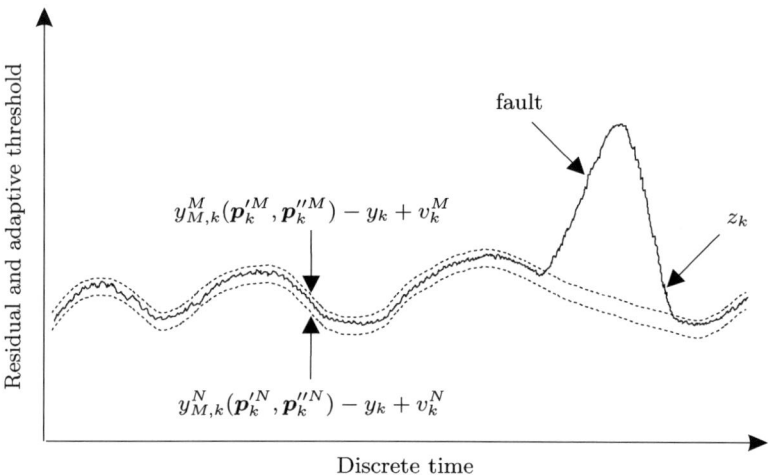

Fig. 3.19 Fault detection with an adaptive threshold developed with the proposed approach

then, as a result of substituting (3.119) into (3.112), the adaptive threshold can be written as

$$y_{M,k}^N \left(p_k'^N, p_k''^N \right) - y_k + v_k^N \leq z_k \leq y_{M,k}^M \left(p_k'^M, p_k''^M \right) - y_k + v_k^M. \quad (3.120)$$

The principle of fault detection with the developed adaptive threshold is shown in Fig. 3.19.

3.2.5 Alternative Robust Fault Detection Procedure: A Backward Detection Test

As can be found in the literature, the procedure proposed in the preceding part of this section (cf. Sect. 3.2.4) is called the *forward test*. Now, an alternative passive robust fault detection test, inspired by the inverse image of an interval function, is introduced. This is why this test named the backward detection test (see [45], where this name is used for describing a very similar approach to the one presented here but in the context of set-membership state estimation). The backward detection test applied to the GMDH neural net consists in checking if there exists $p \in \mathbb{P}$ (cf. (3.90)) such that

$$y_k - v_k^M \leq r_k^T p \leq y_k - v_k^N. \tag{3.121}$$

If there is no $p \in \mathbb{P}$ satisfying (3.121), then a discrepancy between the measured output and the model output is detected and a fault should be indicated. In fact, this test can be perceived as a kind of parameter identification, since the set of parameters that are consistent with the actual set of N_t test measurements is

$$\mathbb{P}_{N_t} = \left\{ p \in \mathbb{R}^{n_p} \mid y_k - v_k^M \leq r_k^T p \leq y_k - v_k^N, k = 1, \ldots, N_t \right\}. \tag{3.122}$$

Then, equivalently, the backward fault detection test consists in checking if

$$\mathbb{P} \cap \mathbb{P}_{N_t} = \emptyset. \tag{3.123}$$

The computation of the parameter set consistent with a given set of N_t measurements

$$\Upsilon = [y_1] \times \cdots \times [y_{N_t}], \tag{3.124}$$

with

$$[y_k] = \left[y_k - v_k^M, y_k - v_k^N \right], \tag{3.125}$$

can be realised with the same bounded-error parameter estimation algorithms as those used for calculating the feasible parameter set or by computing the *inverse image* of the function $g(\cdot)$ describing the GMDH neural network:

$$y_{m,k} = g\left(p_1^{(1)}, \ldots, p_{n_1}^{(1)}, \ldots, p_1^{(L)}, \ldots, p_{n_L}^{(L)} \right), \tag{3.126}$$

where L is the number of layers of the GMDH model and n_l is the number of neurons in the lth neuron. Such an inverse can be described as follows

$$\mathbb{P}_{N_t} = g^{-1}(\Upsilon). \tag{3.127}$$

In [45], the authors proposed proposed an alternative approach called SIVIA (Set Inversion Via Interval Analysis) that computes the inverse image of an interval function using subpavings. This algorithm makes use of the point test

$$t(x) = \left(x \in f^{-1}(\Upsilon) \right), \tag{3.128}$$

associated to the inverse image, which can be easily evaluated by computing $f(x)$ and checking if it belongs to Υ. However, when the dimension of the set to be characterised is high, the computational complexity explodes since SIVIA uses bisection in all directions [45].

Moreover, after computing (3.127), either using bounded-error parameter estimation algorithms or the inverse image of an interval function using SIVIA, the intersection presented in (3.123) should be computed, which is not an easy task in general. When the dimension of the set whose inverse is calculated increases, the calculation time can increase quickly as these algorithms rely on bisections in all directions. Then, by using contractors in combination with bisections the computational burden can be reduced significantly, see [45] for more details. Then, the presented backward test will be implemented using (3.121) with constraint satisfaction algorithms that are polynomial in time. This can be justified by the fact that the use of contractors is computationally less demanding because bisections are only used when required.

Constraint Satisfaction Problem

An Interval Constraint Satisfaction Problem (ICSP) can be formulated as a 3-tuple $\mathbb{H} = (\mathbb{V}, \mathbb{D}, \mathbb{C})$, where $\mathbb{V} = \{\nu_1, \ldots, \nu_n\}$ is a finite set of variables, $\mathbb{D} = \{\nu_1, \ldots, \nu_n\}$ is the set of their domains represented by the closed real intervals, and $\mathbb{C} = \{c_1, \ldots, c_n\}$ is a finite set of constraints relating variables of $\mathbb{V}$. A point solution of $\mathbb{H}$ is a n-tuple $(\tilde{\nu}_1, \ldots, \tilde{\nu}_n) \in \mathbb{V}$ such that all constraints $\mathbb{C}$ are satisfied. The set of all point solutions of $\mathbb{H}$ is denoted by $\mathbb{S}(\mathbb{H})$. This set is called the global solution set. The variable $\nu_i \in \mathbb{V}_i$ is consistent with $\mathbb{H}$ iff

$$\forall \nu_i \in \mathbb{V}_i \quad (\tilde{\nu}_1 \in [\nu_1], \ldots, \tilde{\nu}_n \in [\nu_n]) \,|\, (\tilde{\nu}_1, \ldots, \tilde{\nu}_n) \in \mathbb{S}(\mathbb{H}). \tag{3.129}$$

The principle of algorithms for solving ICSP using local consistency techniques consists essentially in iterating two main operations, domain contraction and propagation, until reaching a stable state. Roughly speaking, if the domain of a variable ν_i is locally contracted with respect to a constraint c_j, then this domain modification is propagated to all the constraints in which v_i occurs leading to the contraction of other variable domains, and so on. Then, the final goal of such a strategy is to contract as much as possible the domains of the variables without loosing any solution by removing inconsistent values through the projection of all constraints. Projecting a constraint with respect to some of its variables consists in computing the smallest interval that contains only consistent values applying a contraction operator.

Backward Test as an ICSP

The fault detection test in (3.121) can be formulated as an ICSP $\mathbb{H} = (\mathbb{V}, \mathbb{D}, \mathbb{C})$ with

$$\mathbb{V} = \{y_1, \ldots, y_k, \boldsymbol{p}\},$$
$$\mathbb{D} = \{[y_1], \ldots, [y_k], \mathbb{P}\},$$
$$\mathbb{C} = \{y_1 = \boldsymbol{r}_1^T \boldsymbol{p}, \ldots, y_k = \boldsymbol{r}_k^T \boldsymbol{p}\}. \qquad (3.130)$$

This problem can be solved by using an ICSP solver as, for example, Interval Peeler developed by the group of Jaulin that employs the principles described in [45] (http://www.ensta-bretagne.fr/jaulin/demo.html). In the case when no solution is found, a fault should be indicated since there is no parameter $\boldsymbol{p} \in \mathbb{P}$ such that (3.121) is satisfied.

In the case of the GMDH neural network, the feasible parameter set $\mathbb{P}$ is not described by an interval parameter vector but by a polygon with vertices $\boldsymbol{p}^i \in V(\mathbb{P})$, $i = 1, \ldots, n_p$ (cf. Fig. 3.14) instead, a slight modification is required in order to get a problem that is compatible with the above-mentioned Interval Peeler. This can be achieved by introducing additional restrictions into (3.130). Each pair of adjacent vertices $\boldsymbol{p}^i$, $\boldsymbol{p}^{i+1}$ introduces a linear restriction of the form

$$f \left(\boldsymbol{p}, p^i, p^{i+1} \right)$$
$$= a_1 \left(\boldsymbol{p}^i, \boldsymbol{p}^{i+1} \right) p_1 + \cdots + a_{n_p} \left(\boldsymbol{p}^i, \boldsymbol{p}^{i+1} \right) p_{n_p} - b \left(\boldsymbol{p}^i, \boldsymbol{p}^{i+1} \right) \leq 0, \quad (3.131)$$

with $\boldsymbol{p} \in \text{hull}(\mathbb{P}) = [\bar{\boldsymbol{p}}, \underline{\boldsymbol{p}}]$, i.e., the minimum interval box containing the feasible parameter set $\mathbb{P}$. Thus, $\text{hull}(\mathbb{P})$ can be easily computed as follows:

$$\bar{p}_i = \min_{\boldsymbol{p} \in V(\mathbb{P})} p_i, \quad i = 1, \ldots, n_p, \qquad (3.132)$$

$$\underline{p}_i = \max_{\boldsymbol{p} \in V(\mathbb{P})} p_i, \quad i = 1, \ldots, n_p. \qquad (3.133)$$

Numerical Example

A simple example in the noise-free context is considered. The main objective is to show how the backward test works and how it outperforms the forward test. Numerous experiments show that this appealing phenomenon appears also when an additive noise is introduced. The example is based on a static interval model defined as follows:

Fig. 3.20 Forward fault
detection test ($k = 1, \ldots, 5$)

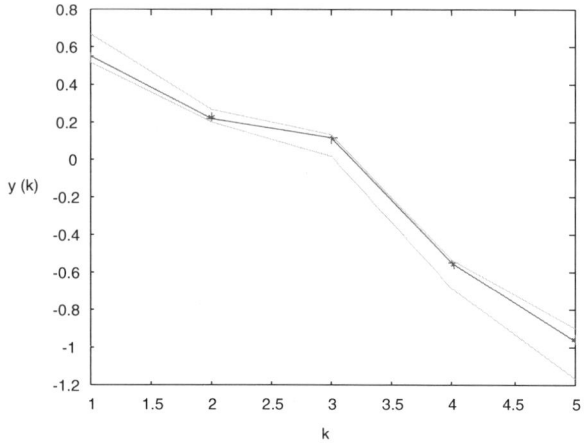

Fig. 3.21 Parameter region
consistent with the forward
test ($k = 1, \ldots, 5$)

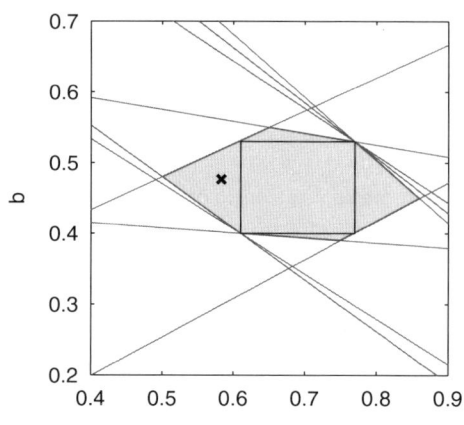

$$y_k = au_{1,k} + bu_{2,k}, \tag{3.134}$$

where $a \in [0.6190,\ 0.7626$ and $b \in [0.3966,\ 0.5402]$ represent the fault-free system
behaviour. Let us consider an abrupt fault scenario related to the changes of a and b
from their nominal values to $a = 0.5829$ and $b = 0.4766$, at time $k = 1$.

The results of the forward detection test are presented in Fig. 3.20. It can be
easily observed that the fault is not detected. As can be seen in Fig. 3.21, the feasible
parameter set (consistent with the data for $k = 1, \ldots, 5$) outerbounds the fault-free
parameter set (represented by the rectangle). The faulty parameter (represented by
the cross) is inside $\mathbb{P}$ but outside the fault-free parameter set. This is the reason why
the forward test does not detect the fault. However, it can be observed in Fig. 3.22
that the fault can be detected for $k = 12$. It can be also seen in Fig. 3.23 that the faulty
parameter (represented by the cross) is outside $\mathbb{P}$, which clearly confirms the result of

Fig. 3.22 Forward fault detection test ($k = 1, \ldots, 5$)

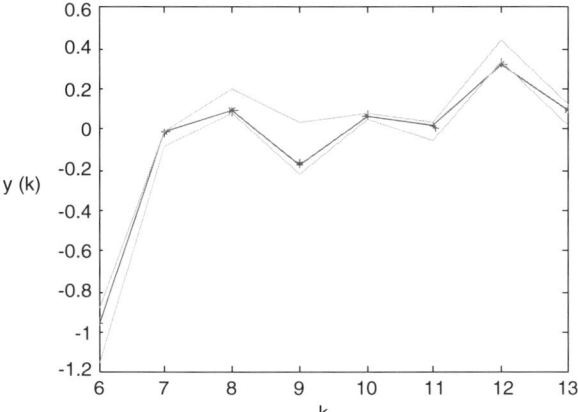

Fig. 3.23 Parameter region consistent with the forward test ($k = 6, \ldots, 13$)

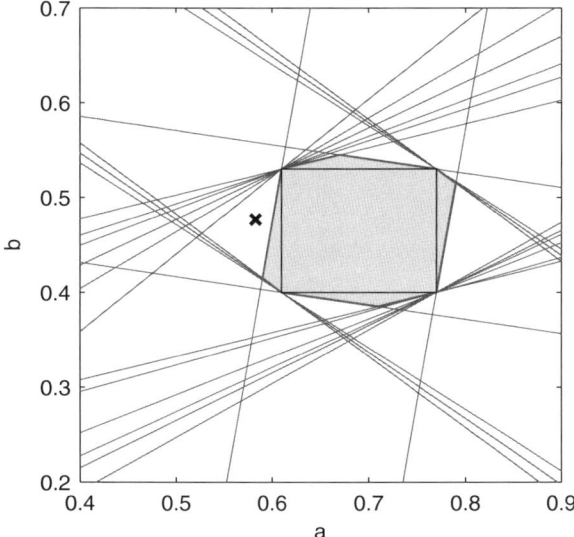

fault detection. On the other hand, using the backward test (3.121) (implemented by 3.130), the set of parameters consistent with the measured data at $k = 1$ is represented by the line presented in Fig. 3.24. Since the line intersects a fault-free parameter set, the fault cannot be detected. Processing the measured data at time instant $k = 2$, one can obtain the second line (Fig. 3.25). It is straightforward to observe that the intersection of the lines corresponding to $k = 1$ and $k = 2$ (cf. (3.130)), results in a point that is outside the fault-free parameter set. Thus, this inconsistency makes it possible to detected the fault being considered (Fig. 3.25).

This simple example clearly shows that the forward test cannot detect some faults even if the faulty parameters are outside the fault-free parameter set. Only after a

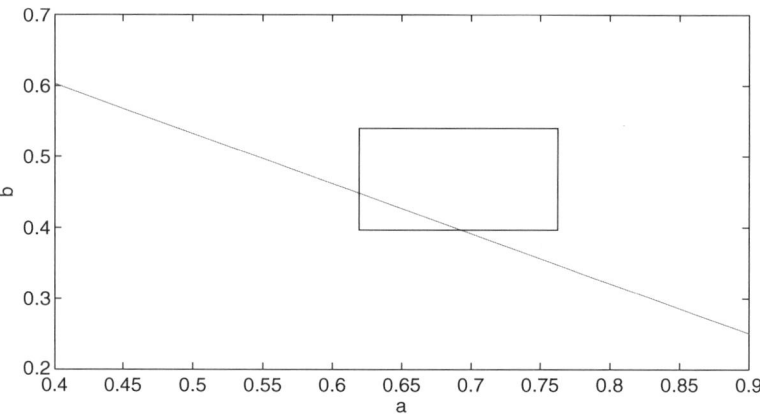

Fig. 3.24 Backward fault detection test at time $k = 1$ (fault not detected)

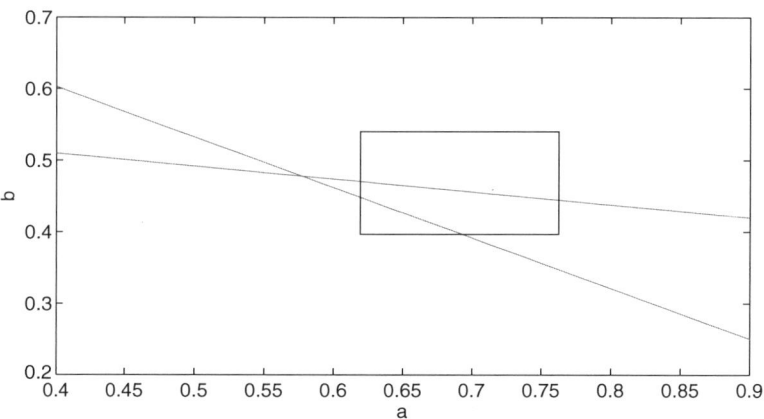

Fig. 3.25 Backward fault detection test at time $k = 2$ (fault detected)

longer time horizon will the output measurement go out of the envelope. On the other hand, the backward test detects the fault much faster, e.g., after two samples in the simple example being considered. Taking into account the above discussion, it can be noticed that the backward test is usually superior over the forward one. Such a superiority results in a significantly shorter fault detection time. This is especially important for industrial applications when a fast fault detection is of a paramount importance, i.e., if the fault is not detected fast enough, then it may cause a sequence of other faults.

3.2.6 *Industrial Application*

The purpose of the present section is to show the effectiveness of the proposed approach in the context of system identification and fault detection with the DAMADICS benchmark. Based on the actuator benchmark definition [27, 46], two GMDH models were designed. These models describe the behaviour of the valve actuator and can be labelled as the juice flow model $F = r_F(X, P_1, P_2, T_1)$ and the servomotor rod displacement model $X = r_X(C_V, P_1, P_2, T_1)$, where r_F and r_X stand for the modelled relation between the inputs F, X and the outputs C_V, X, P_1, P_2, T_1.

The real data used for system identification and the fault detection procedure were collected on 17th November 2001. A detailed description regarding the data and the artificially introduced faults can be found in Table 3.3.

Unfortunately, the data turned out to be sampled too fast. Thus, every 10th value was picked, resulting in the $n_t = 1,000$ training and $n_V = 1,000$ validation data sets. Moreover, the output data should be transformed taking into account the response range of the neuron output. In this section, hyperbolic tangent activation functions were employed and hence this range is $[-1, 1]$. To avoid the saturation of the activation function, the range was further decreased to $[-0.8, 0.8]$. In order to perform data transformation, linear scaling was used. The choice of the neuron structure and the selection method of the neurons in the GMDH network are other important problems of the proposed technique. For that purpose, dynamic neurons [47] and the so-called soft selection method [44] were employed. The dynamics in this neuron are realised by the introduction of a linear dynamic system—an Infinite Impulse Response (IIR) filter. As has previously been mentioned, the quality index of a neuron for the validation data set was defined as

$$Q_V = \frac{1}{n_v} \sum_{k=1}^{n_v} \left| \left(y_{M,k}^M + v_k^M \right) - \left(y_{M,k}^N + v_k^N \right) \right|, \tag{3.135}$$

where $y_{M,k}^M$ and $y_{M,k}^N$ are calculated with (3.98) for the first layer or with (3.108)–(3.109) for the subsequent ones. Table 3.4 presents the evolution of (3.135) for the subsequent layers, i.e., these values are obtained for the best performing neurons in a particular layer. Additionally, for the sake of comparison, the results based on the

Table 3.3 List of data sets

Fault	Range (samples)	Fault/data description
No fault	1–10,000	Training data set
No fault	10,001–20,000	Validation data set
f_{16}	57,475–57,530	Positioner supply pressure drop
f_{17}	53,780–53,794	Unexpected pressure drop across valve
f_{18}	54,600–54,700	Fully or partly opened bypass valves
f_{19}	55,977–56,015	Flow rate sensor fault

Table 3.4 Evolution of $Q_\mathcal{V}$ and $B_\mathcal{V}$ for the subsequent layers

Layer	$r_F(\cdot)$ $Q_\mathcal{V}$	$r_F(\cdot)$ $B_\mathcal{V}$	$r_X(\cdot)$ $Q_\mathcal{V}$	$r_X(\cdot)$ $B_\mathcal{V}$
1	1.5549	0.3925	0.5198	0.0768
2	1.5277	0.3681	0.4914	0.0757
3	1.5047	0.3514	0.4904	0.0762
4	1.4544	0.3334	0.4898	0.0750
5	1.4599	0.3587	0.4909	0.0748

classic quality index [32],

$$B_\mathcal{V} = \frac{1}{n_v} \sum_{k=1}^{n_v} \left| y_k - \hat{y}_k \right|, \tag{3.136}$$

are presented as well.

The results presented in Table 3.4 clearly show that the gradual decrease $Q_\mathcal{V}$ occurs when a new layer is introduced. This follows from the fact that the introduction of a new neuron increases the complexity of the model as well as its modelling abilities. On the other hand, if the model is too complex, then the quality index $Q_\mathcal{V}$ increases. This situation occurs, for both $F = r_F(\cdot)$ and $X = r_X(\cdot)$, when the 5th layer is introduced. This means that GMDH neural networks corresponding to $F = r_F(\cdot)$ and $X = r_X(\cdot)$ should have four layers. From Table 3.4, it can be also seen that the application of the quality index $B_\mathcal{V}$ gives similar results for $F = r_F(\cdot)$, i.e., the same number of layers was selected, whilst for $X = r_X(\cdot)$ it leads to the selection of too simple a structure, i.e., a neural network with only two layers is selected. This implies that the quality index $Q_\mathcal{V}$ makes it possible to obtain a model with a smaller uncertainty. In order to achieve the final structure of $F = r_F(\cdot)$ and $X = r_X(\cdot)$, all unnecessary neurons were removed, leaving only those that are relevant for the computation of the model output. The final structures of GMDH neural networks are presented in Figs. 3.26 and 3.27. From Fig. 3.27, it can be seen that the input variable P_2, was removed during the model development procedure. Nevertheless, the quality index $Q_\mathcal{V}$ achieved a relatively low level. It can be concluded that P_2 has a relatively small influence on the servomotor rod displacement X. This is an example of structural errors that may occur during the selection of neurons in the layer of the GMDH network. On the other hand, the proposed fault detection scheme is robust to such errors. This is because they are taken into account during the calculation of a model output confidence interval.

Figures 3.28 and 3.29 present the modelling abilities of the obtained models $F = r_F(\cdot)$ and $X = r_X(\cdot)$, as well as the corresponding system output confidence interval obtained with the proposed approach for the validation data set.

For the reader's convenience, Fig. 3.30 presents a selected part of Fig. 3.28 for $k = 400$–500 samples. The thick solid line represents the real system output, the thin

Fig. 3.26 Final structure of
$F = r_F(\cdot)$

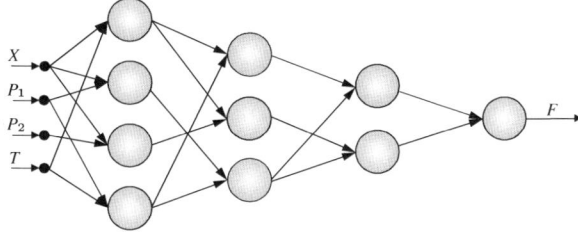

Fig. 3.27 Final structure of
$X = r_X(\cdot)$

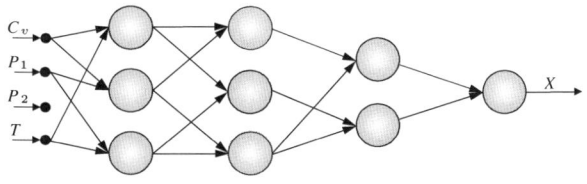

Fig. 3.28 Model and system output as well as the corresponding system output confidence interval for $F = r_F(\cdot)$

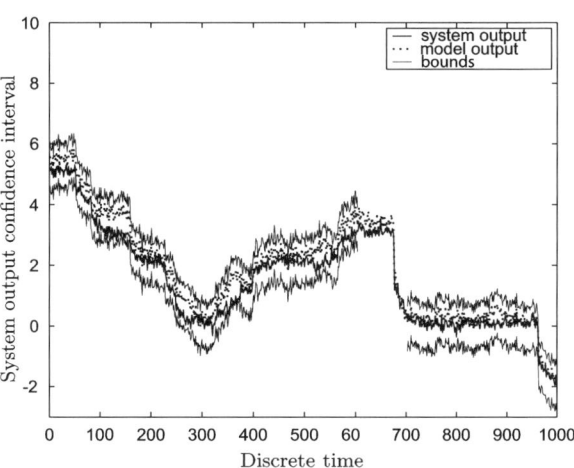

solid lines correspond to the system output confidence interval, and the dashed line is the model output. From Figs. 3.28 and 3.29, it is clear that the system response is contained within system output bounds generated according to the proposed approach. It should be pointed out that these system bounds are designed with the estimated output error bounds. The above estimates were $v_{n_t}^N = -0.8631$ and $v_{n_t}^M = 0.5843$ for $F = r_F(\cdot)$, while $v_{n_t}^N = -0.2523$ and $v_{n_t}^M = 0.2331$ for $X = r_X(\cdot)$.

As has already been mentioned, the quality of the GMDH model can be further improved with the application of the optimisation technique described in [5]. This technique can be perceived as the retraining method for the network. For the valve actuator considered, it was profitable to utilise the retraining technique for the model $F = r_F(\cdot)$. As a result, the quality index (3.136) was decreased from 0.3334 to

Fig. 3.29 Model and system output as well as the corresponding system output confidence interval for $X = r_X(\cdot)$

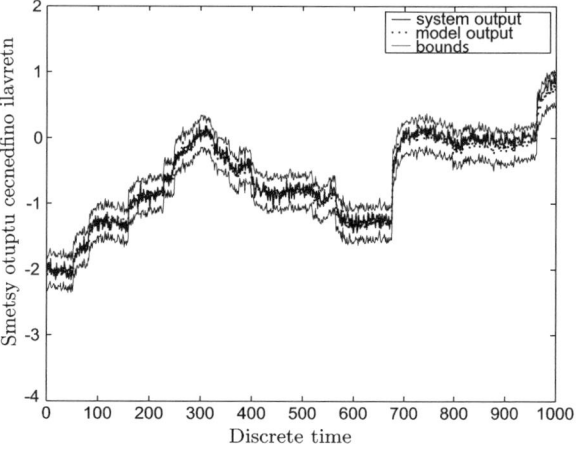

Fig. 3.30 Selected part of Fig. 3.28

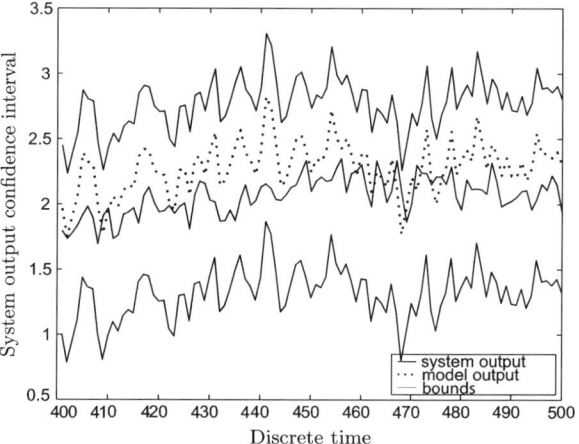

0.2160 (cf. Table 3.4). These results as well as the comparison of Figs. 3.30 and 3.31 justify the need for the retraining technique proposed in [5].

The main objective of this application study was to develop a fault detection scheme for the valve actuator considered. Since both $F = r_F(\cdot)$ and $X = r_X(\cdot)$ were designed with the approach proposed in Sect. 3.2.3, it is possible to employ them for robust fault detection. This task can be realised according to the rules described in Sect. 3.2.4. Figures 3.32, 3.33, 3.34, and 3.35 present the residuals and their bounds for the faulty data.

From these results it can be seen that it is possible to detect all four faults, although the fault f_{18} was detected 18 s after its occurrence. This is caused by the relative insensitivity of the obtained model to this particular fault. The results presented so far were obtained with data from a real system. It should also be pointed out that,

Fig. 3.31 Response of $F = r_F(\cdot)$ after retraining

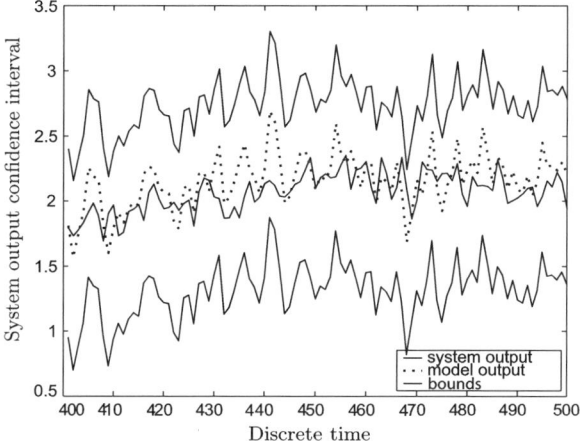

Fig. 3.32 Residual for the fault f_{16}

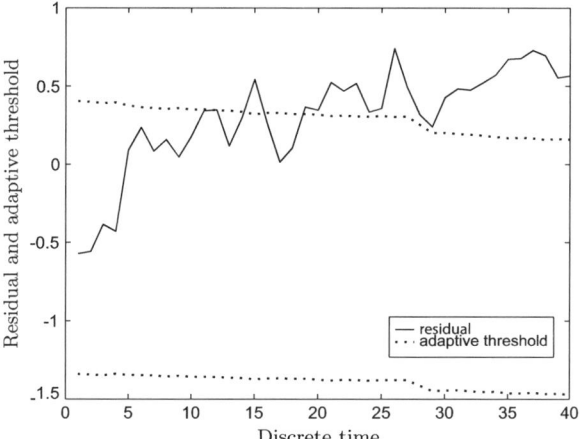

within the framework of the actuator benchmark [46], data for only four general faults f_{16}–f_{19} were available.

In order to provide a more comprehensive and detailed application study of the proposed fault diagnosis scheme, a MATLAB SIMULINK actuator model was employed. This tool makes it possible to generate data for 19 different faults. Table 3.5 shows the results of fault detection. It should be pointed out that both abrupt and incipient faults were considered. As can be seen, the abrupt faults presented in Table 3.5 can be regarded as small, medium and big according to the benchmark description [46]. The notation given in Table 3.5 can be explained as follows: ND means that it is impossible to detect a given fault, D means that it is possible to detect a fault. From the results presented in Table 3.5, it can be seen that it is impossible to detect the faults f_5, f_9 and f_{14}. Moreover, some small and medium faults cannot be detected,

Fig. 3.33 Residual for the
fault f_{17}

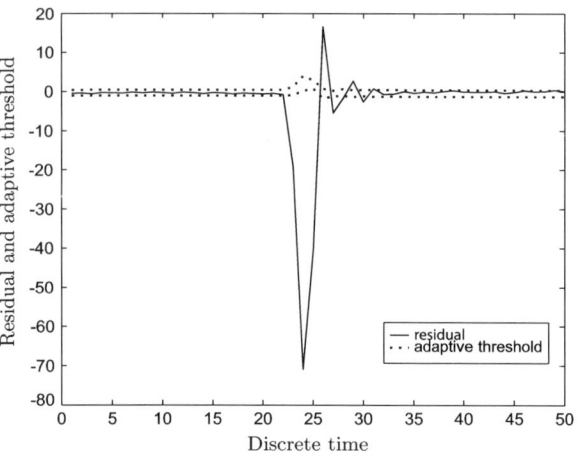

Fig. 3.34 Residual for the
fault f_{18}

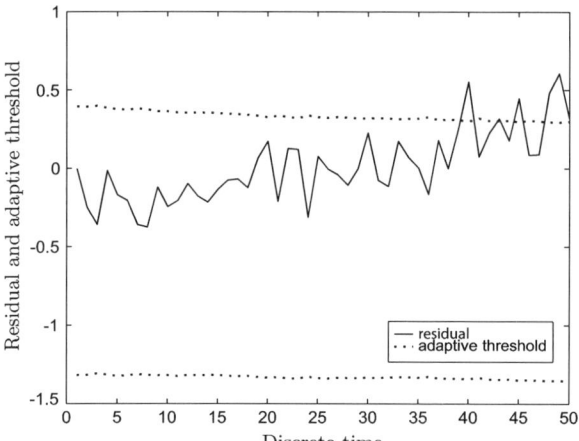

i.e., f_8 and f_{12}. This situation can be explained by the fact that the effect of these
faults is at the same level as that of noise.

Forward Versus Backward Test

The purpose of the subsequent part of this chapter is to perform a comparative case
study regarding the forward and backward fault detection tests. It should be pointed
out that the main objective is to use two fault scenarios clearly indicating the drawbacks and advantages of the approaches considered. In other words, a comprehensive
study regarding the DAMADICS benchmark is beyond the scope of this section.

The first fault scenario considered, named as f_{17} in the DAMADICS benchmark,
consists of an unexpected pressure drop across the valve, which starts at sample

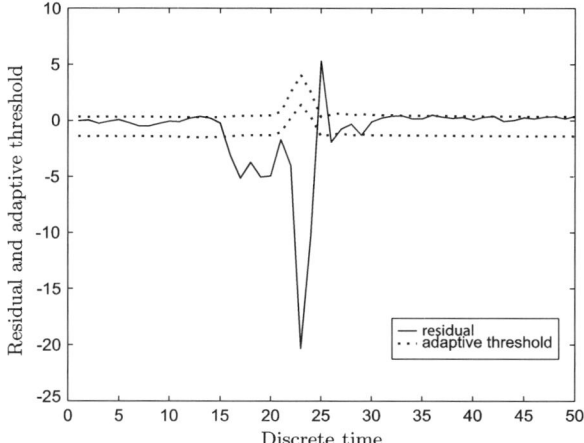

Fig. 3.35 Residual for the fault f_{19}

Table 3.5 Results of fault detection

F	S	M	B	I
f_1	D	D	D	
f_2			D	D
f_3				D
f_4				D
f_5				ND
f_6				D
f_7	D	D	D	
f_8	ND	ND	D	
f_9				ND
f_{10}	D	D	D	
f_{11}			D	D
f_{12}	ND	ND	D_X	
f_{13}	D	D	D	D
f_{14}	ND	ND	ND	
f_{15}			D	
f_{16}	D	D	D	
f_{17}			D	D
f_{18}	D	D	D	D
f_{19}	D	D	D	

S small, M medium, B big, I incipient

$k = 36$ and ends at sample $k = 52$. This real fault scenario was registered at the valve of a servo-actuator. Using the forward test, the fault is detected at sample $k = 38$ (cf. Fig. 3.36) since at this time the measured output leaves the confidence interval (system output uncertainty) provided by the GMDH neural net portrayed in Fig. 3.26, while Fig. 3.37 presents a fault-free part of Fig. 3.36, which clearly exhibits that the

Fig. 3.36 Forward fault
detection test for f_{17} ($k =$
$1, \ldots, 70$)

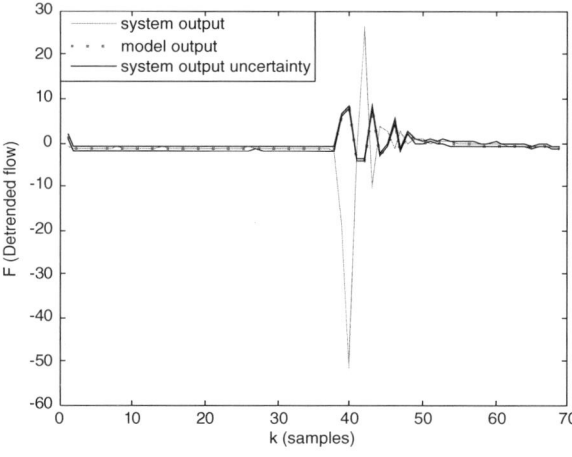

Fig. 3.37 Forward fault
detection test for f_{17} ($k =$
$16, \ldots, 20$)

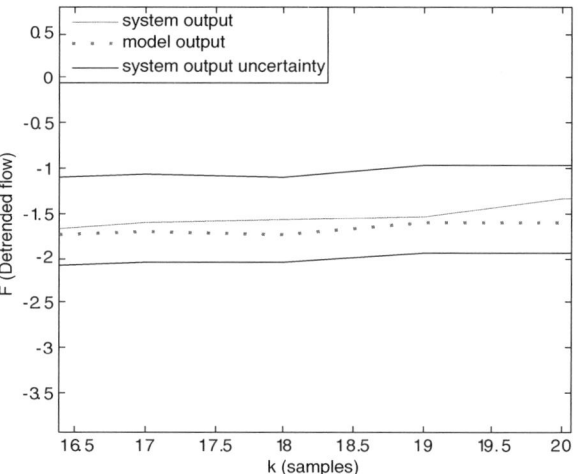

measured output is within the confidence interval. On the other hand, Figs. 3.40
and 3.41 show the results of the backward test. In this case, the fault is detected
at the same sample $k = 38$ (Fig. 3.40) as that for the forward one. Indeed, it can
be shown that the line does not intersect the feasible parameter set obtained in the
training phase (vertices shown with crosses). The set of parameters consistent with
the output envelopes (in a dotted line) also do not contain the parameters consistent
with the measurements. This explains why measurements go outside the envelope.
In Figs. 3.38 and 3.39, the result provided by the backward test is presented when
$k = 37$. In this case, the parameter strip consistent with the measurements intersects
the feasible parameter set, determined in the training phase, and hence the fault is
not detected. The set of parameters consistent with the envelope intersects with the
feasible parameter set. This explains why at that time the envelope is not violated.

Fig. 3.38 Backward fault detection test for f_{17} ($k = 37$)

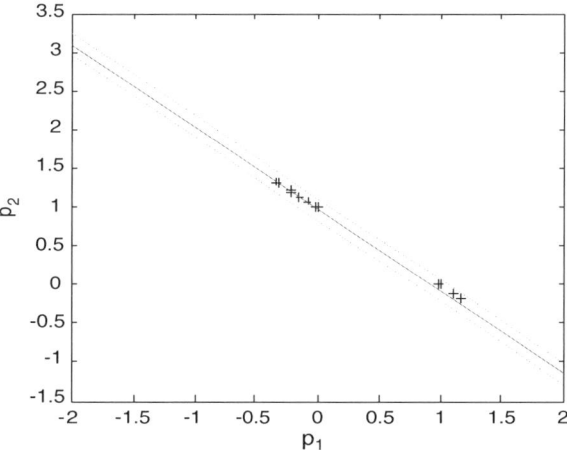

Fig. 3.39 Backward fault detection test for f_{17} ($k = 37$)—zoomed

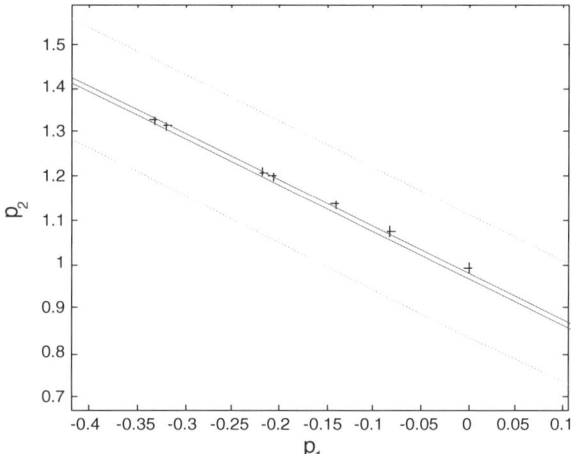

The conclusion is that both forward and backward tests provided the same results. The fault is detected at the same time. This means that it is impossible to show, which of the tests is superior. However, the positive observation is that all of them perform reliably for a given fault, which was obtained with the real valve actuator. The second fault scenario considered, named f_{19} in the DAMADICS benchmark, deals with a flow rate sensor fault. As previously, this is a real fault collected with the valve actuator.

In the case of the forward test, the fault is detected at $k = 32$ (cf. Figs. 3.42 and 3.43). Similarly as for the fault f_{17}, the measured output is inside the confidence interval. This confirms the quality of the obtained neural network and its description of uncertainty. Indeed, this proves that the network possesses good generalisation abilities. On the other hand, Figs. 3.44 and 3.45 present the fault detection result

Fig. 3.40 Backward fault
detection test for f_{17} ($k = 38$)

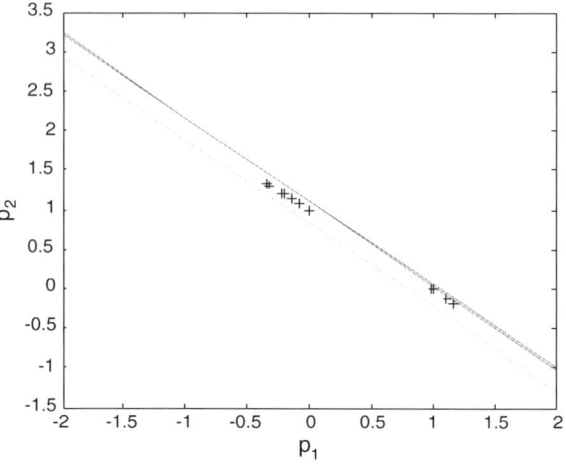

Fig. 3.41 Backward fault
detection test for f_{17} ($k =$
38)—zoomed

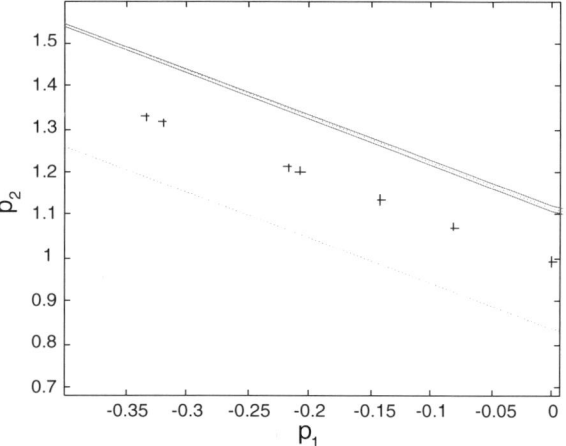

when the backward test is applied. In this case, a measurement at time $k = 28$
enables fault detection. This means that the backward test is superior over the forward one. Note that this superiority may have serious practical consequences. Indeed,
the control action can be performed faster, which decreases the possibility of failure. Finally, it should be noted that both techniques are based on the same feasible
parameter set obtained during the training of the network. The presented results as
well as many experiments with different systems clearly indicate that the backward
test is an attractive alternative to the forward one. As can be observed in the literature, the forward fault detection test based on propagating parameter uncertainty
to the residual or predicted output is predominant. This is, of course, related with
the historical reasons that are strongly settled in statistical theory regarding parameter estimation-based techniques. It should be pointed out that such approaches have

Fig. 3.42 Forward fault detection test for f_{19}

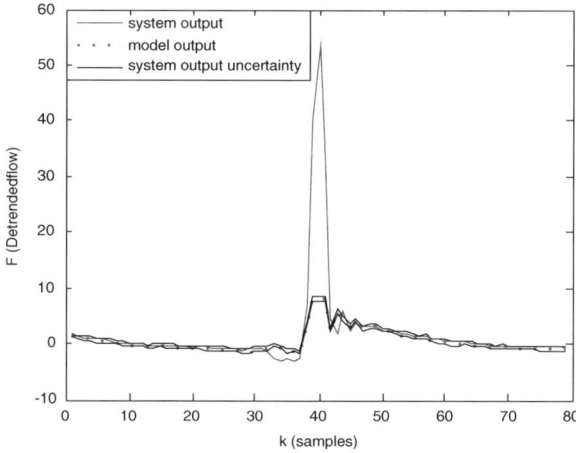

Fig. 3.43 Forward fault detection test for f_{19}—zoomed

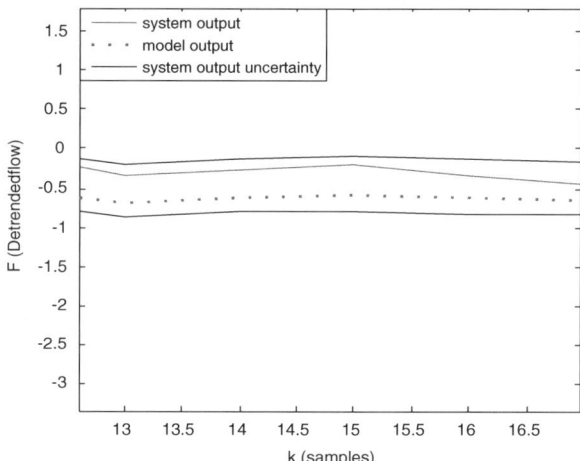

mainly been employed for linear systems. There are, of course, some linearisation-based techniques that extend this technique for non-linear systems. However, their performance is usually very limited. This restricts their practical application. The proposed GMDH-based scheme avoids linearisation while dealing with non-linear systems, which is its unquestionable appeal.

3.3 Concluding Remarks

The portrayed chapter presents two complete design procedures concerning the application of neural networks to robust fault detection. Section 3.1 showed how to settle such a challenging task with a multi-layer perceptron. In particular, it was pictured

Fig. 3.44 Backward fault
detection test for f_{19} ($k = 28$)

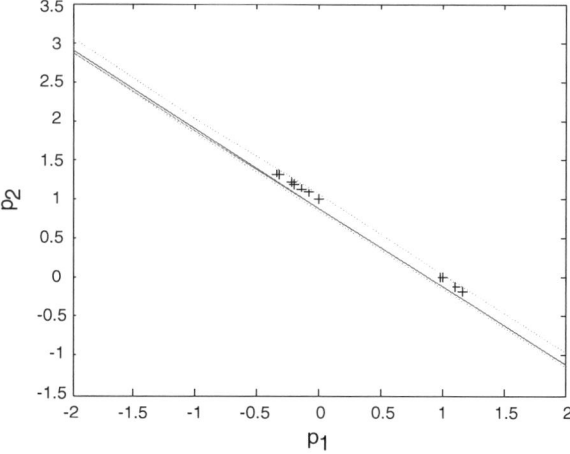

Fig. 3.45 Backward fault
detection test (zoomed) for
f_{19} ($k = 28$)

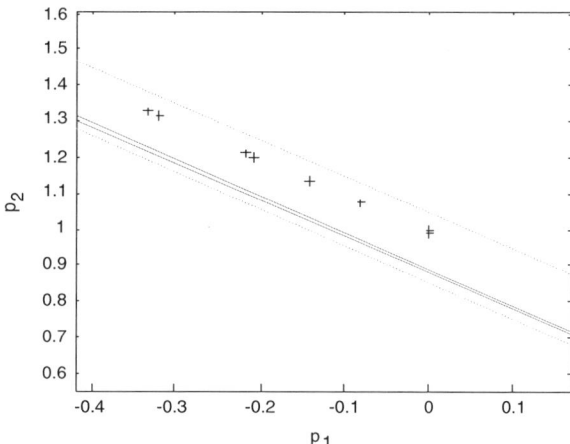

how to describe model uncertainty of the MLP with statistical techniques. Subsequently, an algorithm that can be used for decreasing such a model uncertainty with the use of experimental design theory was presented and described in detail. It was also shown how to use the resulting knowledge about model uncertainty for robust fault detection with the so-called adaptive threshold. The chapter presented numerical examples that show all profits that can be gained while using the proposed algorithms. An industrial application case study concerning the DAMADICS benchmark was also presented. In particular, it was presented how to design an experiment for the MLP being the model of the valve actuator. It was also shown how to use the resulting model and the knowledge about its uncertainty for a robust fault detection study based on a set of selected faults.

A similar task was realised in Sect. 3.2. Starting from a set of input–output measurements of the system, it was shown how to estimate the parameters and the corresponding uncertainty of a neuron via the BEA. The methodology developed for parameter and uncertainty estimation of a neuron makes it possible to formulate an algorithm that allows obtaining a neural network with a relatively small modelling uncertainty. Subsequently, a complete design procedure of a neural network was proposed and carefully described. All the hard computations regarding the design of the GMDH neural network are performed off-line and hence the problem regarding the time-consuming calculations is not of paramount importance. Based on the GMDH neural network, a novel robust fault detection scheme was proposed which supports diagnostic decisions. The scheme is called the forward fault detection test. Moreover, an alternative backward fault detection test was proposed.

Similarly as in Sect. 3.1, the presented approach was tested with the DAMADICS benchmark using both forward and backward approaches.

The experimental results presented in this chapter clearly show all profits that can be gained while using the proposed neural network-based fault detection schemes. It is worth noting that they can be successfully employed instead of the classic techniques, e.g., the described unknown input observers. Indeed, the robustness of the presented fault detection tools makes them useful for solving challenging design problems that arise in engineering practice.

Finally, it should be pointed out that the techniques presented in this chapter are continuously developing towards new and more challenging tasks [48, 49]. The reader is also referred to the alternative approach for designing robust multi-layer perceptron-based fault detection with the bounded error approach [50].

References

1. J. Chen, R.J. Patton, *Robust Model Based Fault Diagnosis for Dynamic Systems* (Kluwer Academic Publishers, London, 1999)
2. K. Patan, M. Witczak, J. Korbicz, Towards robustness in neural network based fault diagnosis. Int. J. Appl. Math. Comput. Sci. **18**(4), 443–454 (2008)
3. M. Witczak, Toward the training of feed-forward neural networks with the D-optimum input sequence. IEEE Trans. Neural Netw. **17**(2), 357–373 (2006)
4. M. Witczak, P. Prętki, Designing neural-network-based fault detection systems with D-optimum experimental conditions. Comput. Assist. Mech. Eng. Sci. **12**(2), 279–291 (2005)
5. M. Witczak, *Modelling and Estimation Strategies for Fault Diagnosis of Non-linear Systems* (Springer, Berlin, 2007)
6. M. Milanese, J. Norton, H. Piet-Lahanier, E. Walter, *Bounding Approaches to System Identification* (Plenum Press, New York, 1996)
7. E. Walter, L. Pronzato, *Identification of Parametric Models from Experimental Data* (Springer, London, 1996)
8. M. Witczak, J. Korbicz, M. Mrugalski, R.J. Patton, A GMDH neural network-based approach to robust fault diagnosis: application to the DAMADICS benchmark problem. Control Eng. Pract. **14**(6), 671–683 (2006)
9. V. Puig, M. Witczak, F. Nejari, J. Quevedo, J. Korbicz, A GMDH neural network-based approach to passive robust fault detection using a constraint satisfaction backward test. Eng. Appl. Artif. Intell. **20**(7), 886–897 (2007)

10. A.C. Atkinson, A.N. Donev, *Optimum Experimental Designs* (Oxford University Press, New York, 1992)
11. M. Witczak, R. Rybski, J. Kaczmarek, Impedance measurement with the D-optimum experimental conditions. IEEE Trans. Instrum. Meas. **58**(8), 2535–2543 (2009)
12. K. Fukumizu, Statistical active learning in multilayer perceptrons. IEEE Trans. Neural Netw. **11**(1), 17–26 (2000)
13. D. Uciński, *Optimal Measurement Methods for Distributed Parameter System Identification.* Systems and Control Series (CRC Press, Boca Raton, 2005)
14. G.C. Goodwin, R.L. Payne, *Dynamic Systems Identification. Experiment Design and Data Analysis* (Acadamic Press, New York, 1977)
15. V.V. Fedorov, P. Hackl, *Model-oriented Design of Experiments* (Springer, New York, 1997)
16. G. Chryssolouris, M. Lee, A. Ramsey, Confidence interval prediction for neural network models. IEEE Trans. Neural Netw. **7**(1), 229–232 (1996)
17. K. Fukumizu, A regularity condition of the information matrix of a multi-layer perceptron network. Neural Netw. **9**(5), 871–879 (1996)
18. K. Fukumizu, Active learning in multilayer perceptrons, in *Advances in Neural Information Processing Systems*, ed. by D.S. Touretzky (MIT Press, Cambridge, 1996), pp. 295–301
19. M.M. Gupta, L. Jin, N. Homma, *Static and Dynamic Neural Networks. From Fundamentals to Advanced Theory* (Wiley, New Jersey, 2003)
20. D. Erdogmus, O. Fontenela-Romero, J.C. Principe, Linear-least-square initialization of multilayer perceptrons through backpropagation of the desired response. IEEE Trans. Neural Netw. **16**(2), 325–337 (2005)
21. E.P.J. Boer, E.M.T. Hendrix, Global optimization problems in optimal design of experiments in regression models. J. Global Optim. **18**(2), 385–398 (2000)
22. J. Korbicz, J. Kościelny, Z. Kowalczuk, W. Cholewa (eds.), *Fault Diagnosis. Models, Artificial Intelligence, Applications* (Springer, Berlin, 2004)
23. L. Pronzato, E. Walter, Eliminating suboptimal local minimizers in nonlinear parameter estimation. Technometrics **43**(4), 434–442 (2001)
24. H.J. Sussmann, Uniqueness of the weights for minimal feedworward nets with a given input-output map. Neural Netw. **5**, 589–593 (1992)
25. V.V. Fedorov, *Theory of Optimal Experiments* (Academic Press, New York, 1972)
26. I. Ford, D.M. Titterington, C.P. Kitsos, Recent advances in nonlinear experimental design. Technometrics **31**(1), 49–60 (1989)
27. DAMADICS. Website of the research training network on Development and application of methods for actuator diagnosis in industrial control systems http://diag.mchtr.pw.edu.pl/damadics, 2004
28. M. Metenidis, M. Witczak, J. Korbicz, A novel genetic programming approach to nonlinear system modelling: application to the DAMADICS benchmark problem. Eng. Appl. Artif. Intell. **17**(4), 363–370 (2004)
29. Papers of the special sessions. DAMADICSI,II,III. In Proc. 5th IFAC Symp. Fault Detection Supervision and Safety of Technical Processes, SAFEPROCESS 2003, Washington DC, USA, 2003
30. M. Witczak, J. Korbicz, Observers and genetic programming in the identification and fault diagnosis of non-linear dynamic systems, in *Fault Diagnosis. Models, Artificial Intelligence, Applications*, ed. by J. Korbicz, J.M. Kościelny , Z. Kowalczuk, W. Cholewa (Springer, Berlin, 2004), pp. 457–509
31. M. Witczak. Identification and Fault Detection of Non-Linear Dynamic Systems. Lecture Notes in Control and Computer Science, Vol. 1 (University of Zielona Góra Press, Zielona Góra, 2003)
32. J.E. Mueller, F. Lemke, *Self-Organising Data Mining* (Libri, Hamburg, 2000)
33. D.T. Pham, L. Xing, *Neural Networks for Identification. Prediction and Control* (Springer, London, 1995)
34. M. Mrugalski. Neural Network Based Modelling of Non-linear Systems in Fault Detection Schemes. PhD thesis, Faculty of Electrical Engineering, Computer Science and Telecommunications, University of Zielona Góra, Zielona Góra (2004)

35. A.G. Ivakhnenko, J.A. Mueller, Self-organizing of nets of active neurons. Syst. Anal. Model. Simul. **20**, 93–106 (1995)
36. T. Soderstrom, P. Stoica, *System Identification* (Prentice-Hall International, Hemel Hempstead, 1989)
37. G. Papadopoulos, P.J. Edwards, A.F. Murray, Confidence estimation methods for neural networks: a practical comparison. IEEE Trans. Neural Netw. **12**(6), 1279–1287 (2001)
38. G.A.F. Seber, C.J. Wild, *Nonlinear Regression* (John Wiley and Sons, New York, 1989)
39. S.H. Mo, J.P. Norton, Fast and robust algorithm to compute exact polytope parameter bounds. Math. Comput. Simul. **32**, 481–493 (1990)
40. P.M. Frank, G. Schreier, E.A. Garcia, Nonlinear observers for fault detection and isolation, in *New Directions in Nonlinear Observer Design*, ed. by H. Nijmeijer, T. Fossen (Springer, Berlin, 1999)
41. T. Clement, S. Gentil, Reformulation of parameter identification with unknown-but-bounded errors. Math. Comput. Simul. **30**(3), 257–270 (1988)
42. R. Fletcher, *Practical Methods of Optimization* (John Wiley and Sons, Chichester, 1981)
43. T. Clement, S. Gentil, Recursive membership set estimation for output-error models. Math. Comput. Simul. **32**(5–6), 505–513 (1990)
44. K. Patan, J. Korbicz, Artificial neural networks in fault diagnosis, in *Fault Diagnosis. Models, Artificial Intelligence, Applications*, ed. J. Korbicz, J.M. Kościelny, Z. Kowalczuk , W. Cholewa (Springer, Berlin, 2004), pp. 333–379
45. L. Jaulin, M. Kieffer, O. Didrit, E. Walter, *Applied Interval Analysis, with Examples in Parameter and State Estimation. Robust Control and Robotics* (Springer, London, 2001)
46. M. Bartyś, R. Patton, M. Syfert, S. Heras, J. Quevedo, Introduction to the DAMADICS actuator FDI study. Control Eng. Pract. **14**(6), 577–596 (2006)
47. M. Mrugalski, M. Witczak, Parameter estimation of dynamic GMDH neural networks with the bounded-error technique. J. Appl. Comput. Sci. **10**(1), 77–90 (2002)
48. M. Mrugalski, M. Witczak, State-space GMDH neural networks for actuator robust fault diagnosis. Adv. Electr. Comput. Eng. **12**(3), 65–72 (2012)
49. M. Mrugalski, An unscented Kalman filter in designing dynamic GMDH neural networks for robust fault detection. Int. J. Appl. Math. Comput. Sci. **23**(1), 157–169 (2013)
50. M. Mrugalski, M. Witczak, J. Korbicz, Confidence estimation of the multi-layer perceptron and its application in fault detection systems. Eng. Appl. Artif. Intell. **21**(8), 895–906 (2008)

Part II
Integrated Fault Diagnosis and Control

Chapter 4
Integrated Fault Diagnosis and Control: Principles and Design Strategies

As has already been mentioned, FTC systems are classified into two distinct classes [1]: passive and active. In passive FTC [2–4], controllers are designed to be robust against a set of predefined faults, therefore there is no need for fault diagnosis, but such a design usually degrades the overall performance. In contrast, active FTC schemes react to faults actively by reconfiguring control actions, and so the system stability and acceptable performance are maintained. To achieve that, the control system relies on FDI [5–8] as well as an accommodation technique [9]. Most of the existing works treat FDI and FTC problems separately. Unfortunately, perfect FDI and fault identification are impossible and hence there always is some inaccuracy related to this process. Thus, there is a need for integrated FDI and FTC schemes for both linear and non-linear systems.

As indicated in the preceding chapters, a number of books have been published in the last decade on the emerging problem of FTC. In particular, the book [10], which is mainly devoted to fault diagnosis and its applications, provides some general rules for hardware redundancy-based FTC. On the contrary, the work [11] introduces the concepts of active and passive FTC. It also investigates the problem of performance and stability of FTC under imperfect fault diagnosis. In particular, the authors consider (under a chain of some, not necessarily easy to satisfy, assumptions) the effect of delayed fault detection and imperfect fault identification but the fault diagnosis scheme is treated separately during the design and no real integration of fault diagnosis and FTC is proposed. FTC is also treated in a very interesting work [12], where the number of practical case studies of FTC is presented, i.e., a winding machine, a three-tank system, and an active suspension system. Unfortunately, in spite of the incontestable appeal of the proposed approaches, neither FTC integrated with fault diagnosis nor a systematic approach to non-linear systems are studied. A particular case of a non-linear aircraft model is studied in [13], but the above-mentioned integration problem is also neglected.

One of the approaches to active FTC is control reconfiguration that creates dependable systems by means of appropriate restructuring of feedback control. It responds to severe component faults that open the control loop by on-line redesigning of the

M. Witczak, *Fault Diagnosis and Fault-Tolerant Control Strategies for Non-Linear Systems*, 119
Lecture Notes in Electrical Engineering 266, DOI: 10.1007/978-3-319-03014-2_4,
© Springer International Publishing Switzerland 2014

controller [9]. It is possible to solve the reconfiguration problem by redesigning a new controller for the faulty system for every isolated fault. An optimal controller can be redesigned with the same optimisation problem as in the nominal case [14, 15], but the redesign step can become too complex for large-scale systems. Also if the controller is a human being, the replacement of the controller for reconfiguration implies the need for strong training efforts.

The approaches presented in this chapter rely on the idea of keeping the nominal controller in the loop and avoiding the complete controller redesign by placing a block between the controller output u and the input of all available actuators u_f. The goal of this block is to provide a signal which has the same effect as the broken actuator would have in the nominal system, therefore masking the fault. This approach can be perceived as a kind of *virtual actuator*. The concept of a virtual actuator has already been introduced in [9, 16–18], where it was demonstrated that it can be applied automatically after the actuator fault has been detected, preventing from serious system failures. It is worth mentioning that the approach presented in this chapter could be extended to sensor faults leading to a kind of virtual sensor thanks to the duality principle (see, e.g., [9] for more details).

Focusing on the previous aim, an integrated design procedure of fault identification and fault-tolerant control schemes is introduced. In particular, the general idea of integrating fault identification and control schemes, which takes into account the fault estimation error, is first presented in a linear context. As a result, the so-called separation principle for the controller and the fault identification scheme is developed. Subsequently, the proposed approach is extended to a class of non-linear system. Similarly to the linear case, it is proven that using a suitable control strategy and a fault identification scheme it is possible to obtain an integrated fault-tolerant control framework, which takes into account the fault identification error. The final part of the chapter presents an illustrative example regarding FTC of a twin-rotor system. Finally, it should be pointed out that the results presented in this chapter are based on [19–23].

4.1 FTC Strategy

The general FTC idea presented in this section was originally developed in [24] for both linear and Takagi–Sugeno fuzzy systems. In order to make the chapter self contained and easy to understand, the main FTC idea is firstly described for linear systems and then suitably extended to non-linear systems in the subsequent part of the chapter.

Let us consider the following reference model:

$$x_{k+1} = Ax_k + Bu_k, \tag{4.1}$$

$$y_{k+1} = Cx_{k+1}, \tag{4.2}$$

where $x_k \in \mathbb{R}^n$ stands for the reference state, $y_k \in \mathbb{R}^m$ is the reference output, and $u_k \in \mathbb{R}^r$ denotes the nominal control input.

Let us also consider a possibly faulty system described by the following equations:

$$x_{f,k+1} = Ax_{f,k} + Bu_{f,k} + Lf_k, \tag{4.3}$$

$$y_{f,k+1} = Cx_{f,k+1}, \tag{4.4}$$

where $x_{f,k} \in \mathbb{R}^n$ stands for the system state, $y_{f,k} \in \mathbb{R}^m$ is the system output, $u_{f,k} \in \mathbb{R}^r$ denotes the system input, $f_k \in \mathbb{R}^s$, $(s \leq m)$ is the fault vector, and L stands for its distribution matrix which is assumed to be known.

The main objective of this chapter is to propose a novel FTC strategy, which can be used for determining the system input $u_{f,k}$ such that

- the control loop for the system (4.3)–(4.4) is stable;
- $x_{f,k}$ converges asymptotically to x_k irrespective of the presence of the fault f_k.

The subsequent part of this section shows development details of the scheme that is able to address such a problem.

The crucial idea is to use the following control strategy:

$$u_{f,k} = -S\hat{f}_k + K_1(x_k - \hat{x}_{f,k}) + u_k, \tag{4.5}$$

where $\hat{f}_k$ is the fault estimate. Note that it is not assumed that $x_{f,k}$ is available, but an estimate $\hat{x}_{f,k}$ can be used instead. Thus, the following problems arise:

- to determine $\hat{f}_k$;
- to design K_1 in such a way that the control loop is stable, i.e., the stabilisation problem.

4.1.1 Fault Identification

Let us assume that the following rank condition is satisfied [8]:

$$\text{rank}(CL) = \text{rank}(L) = s. \tag{4.6}$$

This implies that it is possible to calculate

$$H = (CL)^+ = \left[(CL)^T CL\right]^{-1} (CL)^T. \tag{4.7}$$

By multiplying (4.4) by H and then substituting (4.3), it can be shown that

$$f_k = H(y_{f,k+1} - CAx_{f,k} - CBu_{f,k}). \tag{4.8}$$

Thus, if $\hat{x}_{f,k}$ is used instead of $x_{f,k}$, then the fault estimate is given as follows:

$$\hat{f}_k = H(y_{f,k+1} - CA\hat{x}_{f,k} - CBu_{f,k}), \tag{4.9}$$

and the associated fault estimation error is

$$f_k - \hat{f}_k = -HCA(x_{f,k} - \hat{x}_{f,k}). \tag{4.10}$$

Unfortunately, the problem with practical implementation of (4.9) is that it requires $y_{f,k+1}$ and $u_{f,k}$ to calculate $\hat{f}_k$. Thus, (4.9) cannot be directly used to obtain (4.5). Most of the existing approaches suffer from this problem (see, e.g., [25]), although the authors ignore this difficulty. To solve this issue within the framework of the proposed FTC approach, a one step prediction of the form $\bar{f}_k = \alpha_k \hat{f}_{k-1}$ is employed, where α_k is a diagonal matrix being a design parameter (further remedy to this problem is to be provided in the subsequent chapter). Hence, the practical form of (4.5) boils down to

$$u_{f,k} = -S\bar{f}_k + K_1(x_k - \hat{x}_{f,k}) + u_k. \tag{4.11}$$

In most cases, the matrix α_k is taken as the identity matrix, i.e., the one that corresponds to a one time-step prediction. In the cases where the fault behaves in a linear way, it is possible to design the matrix α_k based on the previous fault changes. In cases where faults change in a non-linear fashion, it is also possible to predict the nature of the faults by using, for example, neural networks (see [19] for a comprehensive study).

4.1.2 Stabilisation Problem

By substituting (4.5) into (4.3), it can be shown that

$$x_{f,k+1} = Ax_{f,k} - BS\hat{f}_k + BK_1(e_k + e_{f,k}) + Bu_k + Lf_k, \tag{4.12}$$

where $e_k = x_k - x_{f,k}$ stands for the tracking error while $e_{f,k} = x_{f,k} - \hat{x}_{f,k}$ is the state estimation error. Let us assume that S satisfies the equality $BS = L$, e.g., $S = I$, for the actuator faults. Thus, $BS = L$ and hence

$$x_{f,k+1} = Ax_{f,k} + L(f_k - \hat{f}_k) + BK_1(e_k + e_{f,k}) + Bu_k. \tag{4.13}$$

Finally, substituting (4.10) into (4.13) and then applying the result into $e_{k+1} = x_{k+1} - x_{f,k+1}$ yield

$$e_{k+1} = (A - BK_1)e_k + (LHCA - BK_1)e_{f,k}. \tag{4.14}$$

4.1.3 Observer Design

As has already been mentioned, the fault estimate (4.9) is obtained based on the state estimate $\hat{x}_{f,k}$. This raises the necessity for observer design. Consequently, by substituting (4.8) into (4.3) it is possible to show that

$$x_{f,k+1} = \bar{A}x_{f,k} + \bar{B}u_{f,k} + \bar{L}y_{f,k+1}, \tag{4.15}$$

where

$$\bar{A} = (I - LHC)A, \quad \bar{B} = (I - LHC)B, \quad \bar{L} = LH.$$

Thus, the observer structure, which can be perceived as an unknown input observer (see, e.g. [8, 26]), is given by

$$\hat{x}_{f,k+1} = \bar{A}\hat{x}_{f,k} + \bar{B}u_{f,k} + \bar{L}y_{f,k+1} + K_2(y_{f,k} - C\hat{x}_{f,k}). \tag{4.16}$$

Finally, the state estimation error can be written as follows:

$$e_{f,k+1} = (\bar{A} - K_2C)e_{f,k}. \tag{4.17}$$

4.1.4 Integrated Design Procedure

The main objective of this section is to sumarise the presented results within an integrated framework for fault identification and fault-tolerant control scheme development. First, let us start with two crucial assumptions:

- the pair $(\bar{A}, C)$ is detectable;
- the pair (A, B) is stabilisable.

Under these assumptions, it is possible to design the matrices K_1 and K_2 in such a way that the extended error

$$\bar{e}_k = \begin{bmatrix} e_k \\ e_{f,k} \end{bmatrix}, \tag{4.18}$$

described by

$$\bar{e}_{k+1} = \begin{bmatrix} A - BK_1 & LHCA - BK_1 \\ 0 & \bar{A} - K_2C \end{bmatrix} \bar{e}_k = A_0 \bar{e}_k, \tag{4.19}$$

converges asymptotically to zero.

Theorem 4.1 *The extended error $\bar{e}_k$ converges asymptotically to zero iff there exist matrices $W \succ 0$, L_1 and $P_2 \succ 0$, L_2 such that*

$$\begin{bmatrix} W & AW - BL_1 \\ WA^T - L_1^T B^T & W \end{bmatrix} \succ 0, \tag{4.20}$$

$$\begin{bmatrix} P_2 & P_2\bar{A} - L_2C \\ \bar{A}^T P_2 - C^T L_2^T & P_2 \end{bmatrix} \succ 0. \tag{4.21}$$

Proof It can be observed from the structure of A_0 in (4.19) that the eigenvalues of the matrix A_0 are the union of those of $A - BK_1$ and $\bar{A} - K_2C$. This clearly indicates that the design of the state feedback and the observer can be carried out independently (separation principle). Thus, it is clear from the Lyapunov theorem that $\bar{e}_k$ converges asymptotically to zero iff there exist matrices $P_1 \succ 0$ and $P_2 \succ 0$ such that following inequalities are satisfied:

$$(A - BK_1)^T P_1 (A - BK_1) - P_1 \prec 0, \tag{4.22}$$

$$(\bar{A} - K_2C)^T P_2 (\bar{A} - K_2C) - P_2 \prec 0. \tag{4.23}$$

By applying the Schur complements, it is possible to show that (4.22)–(4.23) are equivalent to

$$\begin{bmatrix} P_1^{-1} & A - BK_1 \\ A^T - K_1^T B^T & P_1 \end{bmatrix} \succ 0, \tag{4.24}$$

$$\begin{bmatrix} P_2^{-1} & \bar{A} - K_2C \\ \bar{A}^T - C^T K_2^T & P_2 \end{bmatrix} \succ 0. \tag{4.25}$$

By substituting $W = P_1^{-1}$ and then multiplying (4.24) from left and right by $\text{diag}(I, W)$ and (4.25) from left and right by $\text{diag}(P_2, I)$, it can be shown that

$$\begin{bmatrix} W & AW - BK_1W \\ WA^T - WK_1^T B^T & W \end{bmatrix} \succ 0, \tag{4.26}$$

$$\begin{bmatrix} P_2 & P_2\bar{A} - P_2K_2C \\ \bar{A}^T P_2 - C^T K_2^T P_2 & P_2 \end{bmatrix} \succ 0. \tag{4.27}$$

Subsequently, by substituting $L_1 = K_1W$ and $L_2 = P_2K_2$ it is possible to transform (4.26) and (4.27) into (4.20)–(4.21), which completes the proof.

Finally, the design procedure boils down to solving the LMIs (4.20) and (4.21), and then determining $K_1 = L_1W^{-1}$ and $K_2 = P_2^{-1}L_2$.

4.2 Extension to Non-linear Systems

This section shows that the approach presented in Sect. 4.1 can be extended to non-linear systems with the reference model given by

$$x_{k+1} = Ax_k + Bu_k + g(x_k),\tag{4.28}$$
$$y_{k+1} = Cx_{k+1},\tag{4.29}$$

where $g(x)$ is a non-linear function satisfying

$$(g(x_1) - g(x_2))^T (g(x_1) - g(x_2))$$
$$\leq \gamma^2 (x_1 - x_2)^T (x_1 - x_2),\ \forall x_1, x_2 \in \mathbb{X} \subset \mathbb{R}^n,\tag{4.30}$$

where $\gamma > 0$ stands for the Lipschitz constant.

Assuming now that the possibly faulty non-linear system is described by

$$x_{f,k+1} = Ax_{f,k} + Bu_{f,k} + g(x_{f,k}) + Lf_k,\tag{4.31}$$
$$y_{f,k+1} = Cx_{f,k+1},\tag{4.32}$$

let us define

$$s_k = g(x_{f,k}) - g(\hat{x}_{f,k}),\tag{4.33}$$

$$\omega_k = g(x_k) - g(x_{f,k})\tag{4.34}$$

and

$$\gamma_k = g(x_k) - g(\hat{x}_{f,k}).\tag{4.35}$$

Following the same line of reasoning as in Sect. 4.1, it can be shown that the fault estimate is given by

$$\hat{f}_k = H\left(y_{f,k+1} - CA\hat{x}_{f,k} - CBu_{f,k} - Cg(\hat{x}_{f,k})\right),\tag{4.36}$$

and the associated fault estimation error is

$$f_k - \hat{f}_k = -HCA(x_{f,k} - \hat{x}_{f,k}) - HCs_k.\tag{4.37}$$

In contrast to the linear case, the control strategy is given

$$u_{f,k} = -S\hat{f}_k + K_1(x_k - \hat{x}_{f,k}) + K_2\gamma_k + u_k\tag{4.38}$$

Similarly, the observer structure is

$$\hat{x}_{f,k+1} = \bar{A}\hat{x}_{f,k} + \bar{B}u_{f,k} + \bar{L}y_{f,k+1} + \bar{g}\left(\hat{x}_{f,k}\right) + K_3(y_{f,k} - C\hat{x}_{f,k}), \qquad (4.39)$$

with

$$\bar{g}\left(\cdot\right) = (I - LHC)g(\cdot) = Gg(\cdot). \qquad (4.40)$$

Assuming that $S = I$ (actuator faults) and then substituting $L = B$, it can be shown that the e_{k+1} is given as follows:

$$e_{k+1} = A_1 e_k + [\bar{A} - A_1]e_{f,k} + \omega_k + BHCs_k - BK_2\gamma_k, \qquad (4.41)$$

where $A_1 = A - BK_1$.

Finally, setting $K_2 = HC$ yields

$$e_{k+1} = A_1 e_k + [\bar{A} - A_1]e_{f,k} + G\omega_k. \qquad (4.42)$$

Similarly, the state estimation error is given by

$$e_{f,k+1} = A_2 e_{f,k} + Gs_k, \qquad (4.43)$$

where $A_2 = \bar{A} - K_3 C$.

It can be easily observed that the system described by (4.42)–(4.43) is a cascaded non-linear discrete-time system [27, 28]. Observing that $e_{f,k}$ enters linearly into (4.42) and using the results presented in the work [28], the system (4.42)–(4.43) is asymptotically stable only if (4.43) is asymptotically stable and the system

$$e_{k+1} = A_1 e_k + G\omega_k, \qquad (4.44)$$

is asymptotically stable as well. It can be concluded that this property can be perceived as a *separation principle* for the proposed control and fault estimation scheme.

Theorem 4.2 *The tracking error e_k converges asymptotically to zero if there exist matrices $X \succ 0, U \succ 0, N$ and scalars $\bar{\beta} > 0, \bar{\alpha} > 0$ such that*

$$\begin{bmatrix} W & Z^T \\ Z & Y \end{bmatrix} \prec 0, \qquad (4.45)$$

with

$$W = \begin{bmatrix} -X & 0 & 0 & \eta\tilde{A}^T \\ 0 & -U & 0 & \eta XG^T \\ 0 & 0 & U & 0 \\ \eta\tilde{A} & \eta GX & 0 & -\eta X \end{bmatrix}, \qquad (4.46)$$

$$Z = [diag(\gamma X, X, X,), \mathbf{0}_{3n \times n}], \qquad (4.47)$$

$$Y = diag(\bar{\beta}I, -\bar{\beta}I, -\bar{\alpha}I) \tag{4.48}$$

and

$$\begin{bmatrix} -X & X \\ X & -\bar{\alpha}I \end{bmatrix} \prec 0, \tag{4.49}$$

with $\eta = 1 + \gamma^2$ *and* $\tilde{A} = AX - BN$.

Proof The results presented below are in some part inspired by the paper [29]. Let us define the Lyapunov function of the form

$$V_k = e_k^T P e_k + \omega_k^T Q \omega_k, \tag{4.50}$$

with $P \succ 0$ and $Q \succ 0$. Thus, $\Delta V = V_{k+1} - V_k$ can be written as follows:

$$\Delta V = e_{k+1}^T P e_{k+1} + \omega_{k+1}^T Q \omega_{k+1} - e_k^T P e_k - \omega_k^T Q \omega_k. \tag{4.51}$$

By substituting (4.44) into (4.51) it can be shown that

$$\Delta V = e_k^T \left[A_1^T P A_1 - P \right] e_k + \omega_{k+1}^T Q \omega_{k+1} + \omega_k^T G^T P A_1 e_k$$
$$+ e_k^T A_1^T P G \omega_k + \omega_k^T \left[G^T P G - Q \right] \omega_k, \tag{4.52}$$

which can be rewritten as

$$\Delta V = v_k^T \begin{bmatrix} A_1^T P A_1 - P & A_1^T P G & 0 \\ G^T P A_1 & G^T P G - Q & 0 \\ 0 & 0 & Q \end{bmatrix} v_k, \tag{4.53}$$

with $v_k = [e_k^T, \omega_k^T, \omega_{k+1}^T]^T$.

Additionally, from (4.30)

$$\beta \gamma^2 e_k^T e_k - \beta \omega_k^T \omega_k \geq 0, \quad \beta > 0, \tag{4.54}$$

which can be rewritten as follows:

$$v_k^T \begin{bmatrix} \beta \gamma^2 I & 0 & 0 \\ 0 & -\beta I & 0 \\ 0 & 0 & 0 \end{bmatrix} v_k \geq 0. \tag{4.55}$$

Similarly, from (4.30) and by assuming that $P \succ \alpha I$, $\alpha > 0$, it can be shown that

$$\gamma^2 e_{k+1}^T P e_{k+1} - \alpha \omega_{k+1}^T \omega_{k+1} > \alpha \gamma^2 e_{k+1}^T e_{k+1} - \alpha \omega_{k+1}^T \omega_{k+1} \geq 0, \tag{4.56}$$

which can be rewritten as follows:

$$\boldsymbol{v}_k^T \begin{bmatrix} \gamma^2 \boldsymbol{A}_1^T \boldsymbol{P} \boldsymbol{A}_1 & \gamma^2 \boldsymbol{A}_1^T \boldsymbol{P} \boldsymbol{G} & \boldsymbol{0} \\ \gamma^2 \boldsymbol{G}^T \boldsymbol{P} \boldsymbol{A}_1 & \gamma^2 \boldsymbol{G}^T \boldsymbol{P} \boldsymbol{G} & \boldsymbol{0} \\ \boldsymbol{0} & \boldsymbol{0} & -\alpha \boldsymbol{I} \end{bmatrix} \boldsymbol{v}_k \geq 0, \tag{4.57}$$

Consequently, from (4.53)–(4.57) it can be shown that

$$\Delta V \leq \boldsymbol{v}_k^T \boldsymbol{M} \boldsymbol{v}_k, \tag{4.58}$$

where

$$\boldsymbol{M} = \begin{bmatrix} \eta \boldsymbol{A}_1^T \boldsymbol{P} \boldsymbol{A}_1 - \boldsymbol{P} + \beta \gamma^2 \boldsymbol{I} & \eta \boldsymbol{A}_1^T \boldsymbol{P} \boldsymbol{G} & \boldsymbol{0} \\ \eta \boldsymbol{G}^T \boldsymbol{P} \boldsymbol{A}_1 & \eta \boldsymbol{G}^T \boldsymbol{P} \boldsymbol{G} - \boldsymbol{Q} - \beta \boldsymbol{I} & \boldsymbol{0} \\ \boldsymbol{0} & \boldsymbol{0} & \boldsymbol{Q} - \alpha \boldsymbol{I} \end{bmatrix}. \tag{4.59}$$

Thus, if $\boldsymbol{M} \prec \boldsymbol{0}$, then $\Delta V < 0$ and $\boldsymbol{e}_k$ converges asymptotically to zero. Subsequently, using the Schur complements, it is possible to show that $\boldsymbol{M} \prec \boldsymbol{0}$ is equivalent to

$$\begin{bmatrix} -\boldsymbol{P} + \beta \gamma^2 \boldsymbol{I} & \eta \boldsymbol{A}_1^T \boldsymbol{P} & \boldsymbol{0} & \eta \boldsymbol{A}_1^T \boldsymbol{P} \\ \eta \boldsymbol{P} \boldsymbol{A}_1 & -\boldsymbol{Q} - \beta \boldsymbol{I} & \boldsymbol{0} & \eta \boldsymbol{G}^T \boldsymbol{P} \\ \boldsymbol{0} & \boldsymbol{0} & \boldsymbol{Q} - \alpha \boldsymbol{I} & \boldsymbol{0} \\ \eta \boldsymbol{P} \boldsymbol{A}_1 & \eta \boldsymbol{P} \boldsymbol{G} & \boldsymbol{0} & -\eta \boldsymbol{P} \end{bmatrix} \prec \boldsymbol{0}. \tag{4.60}$$

Multiplying (4.60) from left and right by

$$\text{diag}(\boldsymbol{P}^{-1}, \boldsymbol{P}^{-1}, \boldsymbol{P}^{-1}, \boldsymbol{P}^{-1}) \tag{4.61}$$

and then substituting $\boldsymbol{X} = \boldsymbol{P}^{-1}$, $\boldsymbol{U} = \boldsymbol{X} \boldsymbol{Q} \boldsymbol{X}$, and again using the Schur complements (with $\bar{\beta} = \beta^{-1}$ and $\bar{\alpha} = \alpha^{-1}$), it is possible to show that $\boldsymbol{M} \prec \boldsymbol{0}$ is equivalent to (4.45). It should also be noted that $\boldsymbol{P} \succ \alpha \boldsymbol{I}$ is equivalent to (4.49).

Finally, the design procedure boils down to solving (4.45)–(4.49) and then calculating $\boldsymbol{K}_1 = \boldsymbol{N} \boldsymbol{X}^{-1}$.

Theorem 4.3 *The state estimation error $\boldsymbol{e}_{f,k}$ converges asymptotically to zero if there exist scalars $\alpha > 0$, $\beta > 0$ and matrices $\boldsymbol{P} \succ \alpha \boldsymbol{I}$, $\boldsymbol{Q} \succ \boldsymbol{0}$ such that*

$$\begin{bmatrix} -\boldsymbol{P} + \beta \gamma^2 \boldsymbol{I} & \eta \tilde{\boldsymbol{A}}^T & \boldsymbol{0} & \eta \tilde{\boldsymbol{A}}^T \\ \eta \tilde{\boldsymbol{A}} & -\boldsymbol{Q} - \beta \boldsymbol{I} & \boldsymbol{0} & \boldsymbol{G}^T \boldsymbol{P} \\ \boldsymbol{0} & \boldsymbol{0} & \boldsymbol{Q} - \alpha \boldsymbol{I} & \boldsymbol{0} \\ \eta \tilde{\boldsymbol{A}} & \boldsymbol{P} \boldsymbol{G} & \boldsymbol{0} & -\eta \boldsymbol{P} \end{bmatrix} \prec \boldsymbol{0}, \tag{4.62}$$

with $\eta = 1 + \gamma^2$ and $\tilde{\boldsymbol{A}} = \boldsymbol{P} \bar{\boldsymbol{A}} - \boldsymbol{N}_2 \boldsymbol{C}$.

Proof The proof is inspired by [29]. The state estimation error (4.43) for $\boldsymbol{f}_k = \boldsymbol{0}$ can be written as

$$e_{k+1} = (\bar{A} - K_3 C)e_k + G\left(g(x_k) - g(\hat{x}_k)\right) = A_1 e_k + Gs_k. \tag{4.63}$$

Let us define the Lyapunov function of the form

$$V_k = e_k^T P e_k + s_k^T Q s_k, \tag{4.64}$$

with $P \succ 0$ and $Q \succ 0$. Thus, $\Delta V = V_{k+1} - V_k$ can be written as follows:

$$\Delta V = e_{k+1}^T P e_{k+1} + s_{k+1}^T Q s_{k+1} - e_k^T P e_k - s_k^T Q s_k, \tag{4.65}$$

which can be rewritten as

$$\Delta V = v_k^T \begin{bmatrix} A_1^T P A_1 - P & A_1^T P G & 0 \\ G^T P A_1 & G^T P G - Q & 0 \\ 0 & 0 & Q \end{bmatrix} v_k, \tag{4.66}$$

with $v_k = [e_k^T, s_k^T, s_{k+1}^T]^T$.

Additionally, from (4.30),

$$\beta\gamma^2 e_k^T e_k - \beta s_k^T s_k \geq 0, \quad \beta > 0, \tag{4.67}$$

which can be rewritten as

$$v_k^T \begin{bmatrix} \beta\gamma^2 I & 0 & 0 \\ 0 & -\beta I & 0 \\ 0 & 0 & 0 \end{bmatrix} v_k \geq 0. \tag{4.68}$$

Similarly, from (4.30) and by assuming that $P \succ \alpha I$, $\alpha > 0$, it can be shown that

$$\gamma^2 e_{k+1}^T P e_{k+1} - \alpha s_{k+1}^T s_{k+1} > \alpha\gamma^2 e_{k+1}^T e_{k+1} - \alpha s_{k+1}^T s_{k+1} \geq 0, \tag{4.69}$$

which can be rewritten as

$$v_k^T \begin{bmatrix} \gamma^2 A_1^T P A_1 & \gamma^2 A_1^T P G & 0 \\ \gamma^2 G^T P A_1 & \gamma^2 G^T P G & 0 \\ 0 & 0 & -\alpha I \end{bmatrix} v_k \geq 0. \tag{4.70}$$

Consequently, from (4.66)–(4.70) it can be shown that

$$\Delta V \leq v_k^T M v_k, \tag{4.71}$$

where

$$M = \begin{bmatrix} \eta A_1^T P A_1 - P + \beta \gamma^2 I & \eta A_1^T P G & 0 \\ \eta G^T P A_1 & \eta G^T P G - Q - \beta I & 0 \\ 0 & 0 & Q - \alpha I \end{bmatrix}. \tag{4.72}$$

Thus, if $M \prec 0$, then $\Delta V < 0$ and e_k converges asymptotically to zero. Subsequently, using the Schur complements, it is possible to show that $M \prec 0$ is equivalent to

$$\begin{bmatrix} -P + \beta \gamma^2 I & \eta A_1^T P & 0 & \eta A_1^T P \\ \eta P A_1 & -Q - \beta I & 0 & \eta G^T P \\ 0 & 0 & Q - \alpha I & 0 \\ \eta P A_1 & \eta P G & 0 & -\eta P \end{bmatrix} \prec 0. \tag{4.73}$$

Substituting $\tilde{A} = PA_1 = P\bar{A} - PK_3 C = P\bar{A} - N_2 C$ into (4.73) gives (4.62), which completes the proof.

Finally, the design procedure boils down to solving (4.62) and then calculating $K_3 = P^{-1} N_2$.

4.3 Constrained State Estimation

This section presents a straightforward approach that can be used when the state is described by the following bounded state set:

$$\mathbb{X} = \{x : Dx \leq b, x \in \mathbb{R}^n\}, \tag{4.74}$$

with $D \in \mathbb{R}^{c \times n}$, $c \leq n$ being a full-rank matrix. Let us define the following:

$\hat{x}_k$: an unconstrained state estimate obtained with (4.39),
$\bar{x}_k$: a constrained state estimate obtained by projecting $\hat{x}_k$ onto $\mathbb{X}$.

Projection onto $\mathbb{X}$ boils down to solving the following constrained quadratic programming problem:

$$\bar{x}_k = \arg \min_{x_k \in \mathbb{R}^n} (x_k - \hat{x}_x)^T (x_k - \hat{x}_x),$$

$$\text{subject to } Dx_k \leq b. \tag{4.75}$$

There is a number of algorithms that can be applied to solve (4.75), and most of them can be classified as the so-called active set methods.

Let us assume that i of c constraints are active for $\hat{x}_k$, and let us denote by $\bar{D}$ and $\bar{b}$ the i rows of D and i elements of b corresponding to the active constraints. Thus, the problem (4.75) can be reformulated as

$$\bar{x}_k = \arg \min_{x_k \in \mathbb{R}^n} (x_k - \hat{x}_x)^T (x_k - \hat{x}_x),$$

$$\text{subject to } \bar{D}x_k = \bar{b}, \tag{4.76}$$

and its solution is given by

$$\bar{x}_k = \hat{x}_k - \bar{D}^T \left[\bar{D}\bar{D}^T \right]^{-1} \left(\bar{D}\hat{x}_k - \bar{b} \right). \tag{4.77}$$

The appealing property of the constrained state estimate is given by the following theorem.

Theorem 4.4 *Simon and Chia* [30] *The constrained unknown input observer estimate $\bar{x}_k$ given by (4.77) satisfies the inequality*

$$\|x_k - \bar{x}_k\|_2 \leq \|x_k - \hat{x}_k\|_2, \tag{4.78}$$

where $\hat{x}_k$ is the unconstrained state estimate obtained with the unknown input observer (4.39).

4.3.1 Complete Design Procedure

The proposed FTC approach is illustrated in Fig. 4.1. The scheme is composed of four parts:

- *Possibly faulty system* described by (4.31) and (4.32);
- *Reference model* described by (4.28) and (4.29);
- *State estimation and fault identification* described by (4.36) and (4.39), where K_3 is designed by solving (4.62) and then calculating $K_3 = P^{-1}N_2$;
- *Fault-tolerant controller* described by (4.38) where K_1 is designed by solving (4.45)–(4.49) and then calculating $K_1 = NX^{-1}$ while $K_2 = HC$.

Fig. 4.1 Complete FTC scheme

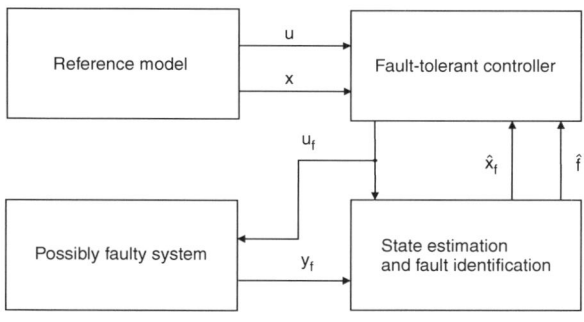

As can be noticed, the complete design procedure involves the solution of a set of linear matrix inequalities. It is also necessary to underline that there is no switching mechanism that changes the control law when a fault is detected and isolated. On the contrary, an integrated scheme is proposed and its convergence is proven within the framework of theorems stated in the preceding part of this chapter. Thus, fault detection and isolation are not necessary in the proposed approach, although it is possible to be done by using the fault estimate (obtained by the module *State estimation and fault identification* in Fig. 4.1). Indeed, the fault can be detected when at least one $\hat{f}_i$ exceeds a predefined threshold δ_i,

$$|\hat{f}_i| > \delta_i, \quad i = 1, \ldots, r. \tag{4.79}$$

Similarly, the ith actuator fault can be isolated when $|\hat{f}_i| > \delta_i$, where δ_i stands for a predefined threshold.

4.4 Application Example

4.4.1 Description of the Twin-Rotor MIMO System

The TRMS is a laboratory setup developed by Feedback Instruments Limited for control experiments.[1] The system is perceived as a challenging engineering problem due to its high non-linearity, cross-coupling between its two axes, and inaccessibility of some of its states through the measurements. The TRMS mechanical unit has two rotors placed on a beam together with a counterbalance whose arm, with a weight at its end, is fixed to the beam at the pivot and determines a stable equilibrium position (Fig. 4.2). The TRMS can rotate freely both in the horizontal and vertical planes by changing the input voltage of the DC motors that drive the rotational speed of the (tail and main) rotors.

The system input vector is $\boldsymbol{u}_k = [U_{h,k}, \ U_{v,k}]^T$, where U_h and U_v are the input voltages of the tail and main motor, respectively. The system states are $\boldsymbol{x}_k = [i_{ah,k}, \omega_{h,k}, \Omega_{h,k}, \theta_{h,k}, i_{av,k}, \omega_{v,k}, \Omega_{v,k}, \theta_{v,k}]^T$, where $i_{ah/av}$ is the armature current of tail/main rotor, $\omega_{h/v}$ is the rotational velocity of the tail/main rotor, $\Omega_{h/v}$ is the angular velocity around the horizontal/vertical axis, and $\theta_{h/v}$ is the azimuth/pitch angle of beam.

---

[1] http://www.feedback-group.com/product/twin-rotor-mimo-6056

Fig. 4.2 Components of the twin-rotor MIMO system

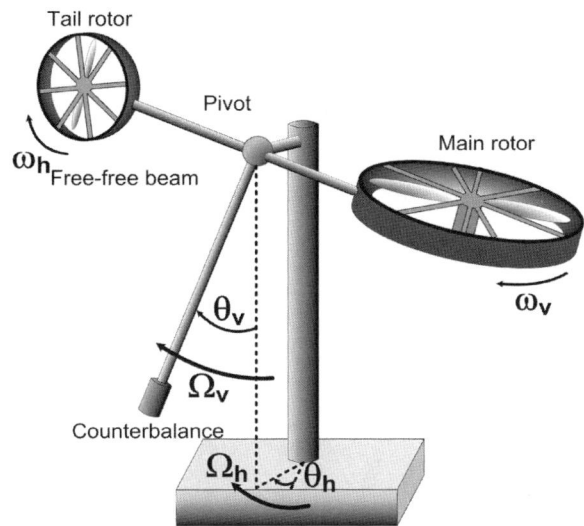

4.4.2 Non-linear Reference Model of Twin-Rotor MIMO System

The non-linear dynamic model of the TRMS supplied by the manufacturer does not represent accurately the system dynamics because some of the effective forces were not taken into account. Alternatively, an accurate non-linear model proposed in [31], including all acting forces, is used instead. This improved non-linear model of the TRMS can be expressed as follows:

$$\frac{di_{ah}}{dt} = -\frac{R_{ah}}{L_{ah}} i_{ah} - \frac{k_{ah}\varphi_h}{L_{ah}} \omega_h + \frac{k_1}{L_{ah}} U_h, \tag{4.80}$$

$$\frac{d\omega_h}{dt} = \frac{k_{ah}\varphi_h}{J_{tr}} i_{ah} - \frac{B_{tr}}{J_{tr}} \omega_h - \frac{f_1(\omega_h)\omega_h^2}{J_{tr}}, \tag{4.81}$$

$$\frac{d\Omega_h}{dt} = \frac{l_t f_1(\omega_h)\omega_h^2 \cos^2\theta_v - k_{oh}\Omega_h - f_3(\theta_h)\theta_h}{D\cos^2\theta_v + E\sin^2\theta_v + F}$$

$$+ \frac{k_m \omega_v \sin\theta_v \Omega_v \left((D - 2E)\cos^2\theta_v - E\sin^2\theta_v - F\right)}{\left(D\cos^2\theta_v + E\sin^2\theta_v + F\right)^2}$$

$$+ \frac{k_m \cos\theta_v \left(k_{av}\varphi_v i_{av} - B_{mr}\omega_v - f_4(\omega_v)\omega_v^2\right)}{\left(D\cos^2\theta_v + E\sin^2\theta_v + F\right) J_{mr}}, \tag{4.82}$$

$$\frac{d\theta_h}{dt} = \Omega_h, \tag{4.83}$$

$$\frac{di_{av}}{dt} = -\frac{R_{av}}{L_{av}} i_{av} - \frac{k_{av}\varphi_v}{L_{av}} \omega_v + \frac{k_2}{L_{av}} U_v, \tag{4.84}$$

$$\frac{d\omega_v}{dt} = \frac{k_{av}\varphi_v}{J_{mr}} i_{av} - \frac{B_{mr}}{J_{mr}} \omega_v - \frac{f_4(\omega_v)\omega_v^2}{J_{mr}}, \tag{4.85}$$

$$\frac{d\Omega_v}{dt} = \frac{l_m f_5(\omega_v)\omega_v^2 + k_g \Omega_h f_5(\omega_v)\omega_v^2 \cos\theta_v - k_{ov}\Omega_v}{J_v}$$

$$+ \frac{g\left((A-B)\cos\theta_v - C\sin\theta_v\right) - 0.5\Omega_v^2 H \sin(2\theta_v)}{J_v}$$

$$+ \frac{k_t \left(k_{ah}\varphi_h i_{ah} - B_{tr}\omega_h - f_1(\omega_h)\omega_h^2\right)}{J_v J_{tr}}, \tag{4.86}$$

$$\frac{d\theta_v}{dt} = \Omega_v, \tag{4.87}$$

where the values of the parameters in (4.80)–(4.87) can be found in [31].

The non-linear model (4.80)–(4.87) was discretised using the zero-order hold and sampling time $T_s = 0.005\,\text{s}$, such that it can be expressed as (4.28)–(4.29) around an equilibrium point x_{eq} as follows:

$$x_{k+1} = A(x_{eq})x_k + B(x_{eq})u_k + g\left(x_{eq}, x_k\right), \tag{4.88}$$

$$y_{k+1} = Cx_k, \tag{4.89}$$

where $A(x_{eq})$ and $B(x_{eq})$ are the frozen system matrices at the equilibrium point. The non-linear function $g\left(x_{eq}, x_k\right)$ is defined as

$$g\left(x_k, x_{eq}\right) = (A(x_k) - A(x_{eq}))x_k, \tag{4.90}$$

where $x_{eq} = [i_{ah,eq},\ \omega_{h,eq},\ \Omega_{h,eq},\ \theta_{h,eq},\ i_{av,eq},\ \omega_{v,eq},\ \Omega_{v,eq},\ \theta_{v,eq}]^T$ are the state variables at the equilibrium point, $u_k = [U_{h,k},\ U_{v,k}]^T$ are the system inputs, and $y_k = [i_{ah,k},\ \theta_{h,k},\ i_{av,k},\ \theta_{v,k}]^T$ are the system outputs. The non-linear equation (4.90) satisfies (4.30) with $\gamma = 10$.

By solving (4.45)–(4.49), it is possible to find the gain K_1:

$$K_1 = \begin{bmatrix} -0.0208 & 0.2308 & 1.13\cdot10^{-6} & 0 & -1.11\cdot10^{-6} & 0 & 1.17\cdot10^{-4} & 0 \\ -1.68\cdot10^{-6} & -1.27\cdot10^{-6} & 1.31\cdot10^{-3} & 0 & -0.2215 & 0.3663 & 2.02\cdot10^{-5} & 0 \end{bmatrix}, \tag{4.91}$$

On the other hand, according to the above-described procedure, it is possible to obtain the gain K_2:

$$K_2 = \begin{bmatrix} 1.2404 & 0 & 0 & 0 & 0 & 0 & 0 & 0 \\ 0 & 0 & 0 & 0 & 0.9422 & 0 & 0 & 0 \end{bmatrix}. \tag{4.92}$$

Finally, the gain K_3 is obtained by solving (4.62)

$$K_3 = \begin{bmatrix} 3.01E-9 & -1.95E-8 & 0 & -3.83E-6 \\ 0 & -1.01E-7 & 2.03E-4 & -1.15E-3 \\ -4.38E-4 & 2.56E-2 & -0.0285 & 1.49E-4 \\ 3.16E-6 & 1.0001 & -1.43E-4 & 2.85E-7 \\ 0 & -3.95E-6 & 5.64E-10 & -2.69E-8 \\ 8.23E-4 & -3.60E-5 & -8.0892 & 4.03E-7 \\ -0.1516 & 6.06E-5 & -2.36E-4 & 1.50E-3 \\ -7.60E-4 & 1.69E-7 & 2.84E-7 & 1.0001 \end{bmatrix}. \tag{4.93}$$

As can be observed, the TRMS has two inputs, which are two voltages, i.e., the main rotor and tail rotor ones. Unfortunately, the TRMS platform has no possibility to introduce some actuator faults related to the main and tail DC motors. Thus, the only way to simulating the faults is to increase (decrease) the voltages feeding the motors. This means that the voltages coming from the controller will be reduced before they reach the DC motors. In this way a drop in DC motor performance will be simulated.

After the application of the design procedure introduced in Sect. 4.2, the FTC scheme is implemented according to Fig. 4.1 and tested within several fault scenarios.

4.4.3 Fault Scenario 1

In the first fault scenario, a fault in the TRMS tail rotor is applied as follows:

$$f_h = \begin{cases} 0\text{V}, & \text{for } t < 20\text{s} \\ -0.2\text{V}, & \text{for } t \geq 20\text{s} \end{cases} \tag{4.94}$$

at the equilibrium point $\theta_{h,ref} = 0.4$ rad and $\theta_{v,ref} = 0.1$ rad. Figure 4.3a presents the response of the azimuth angle of the beam while Fig. 4.3b shows the pitch angle of the beam with (solid line) and without (dashed line) the proposed FTC scheme (4.38). Notice that the TRMS is stabilised in spite of the actuator fault (4.94). However, it can be noticed that without the proposed FTC scheme the TRMS cannot follow the

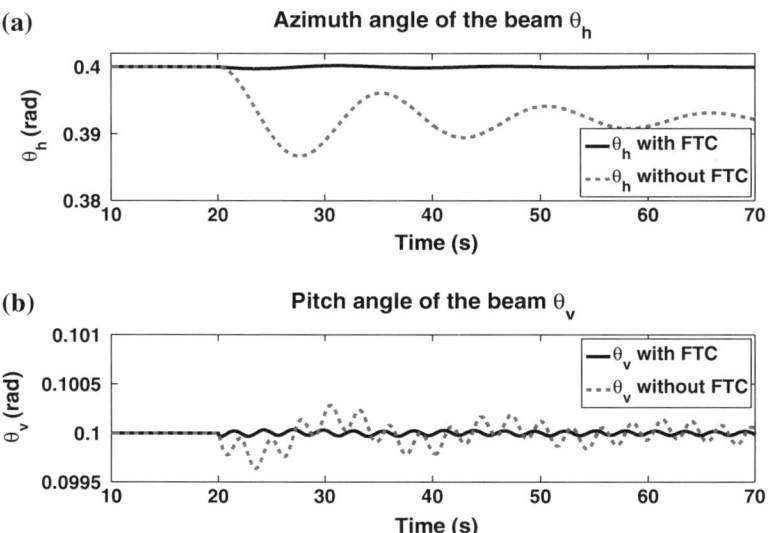

Fig. 4.3 Fault scenario 1: **a** azimuth angle of the beam (*horizontal position*), **b** pitch angle of the beam (*vertical position*)

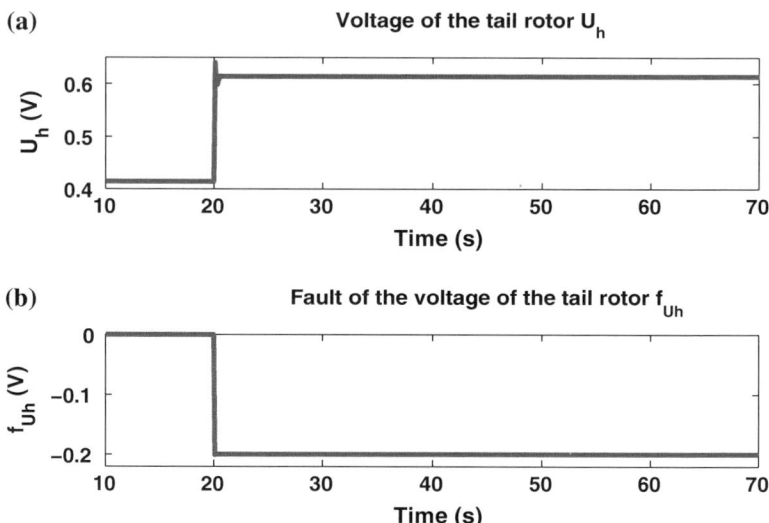

(a) Voltage of the tail rotor U_h

(b) Fault of the voltage of the tail rotor f_{Uh}

Fig. 4.4 Fault scenario 1: **a** voltage of the tail rotor U_h, **b** estimated fault of the voltage of the tail rotor U_h

reference. This is not the case when the FTC scheme is used. Figure 4.4a shows the control action $U_{h,k}$ provided by the FTC scheme (4.38). It can be seen that the control action changes to compensate the actuator fault (4.94) masking the fault effect to the TRMS. Finally, Fig. 4.4b presents the estimated fault provided by (4.36). Notice that the actuator fault is estimated with very high accuracy (see the fault definition (4.94)).

4.4.4 Fault Scenario 2

In this fault scenario the following fault is applied to the main rotor:

$$f_v = \begin{cases} 0\text{V}, & \text{for } t < 20\text{s} \\ -0.3\text{V}, & \text{for } t \geq 20\text{s} \end{cases} \tag{4.95}$$

at the equilibrium point $\theta_{h,ref} = 0.4$ rad and $\theta_{v,ref} = 0.1$ rad.

Figures 4.5a and 4.5b present the response of the azimuth and pitch angle of the beam, respectively. Both angles are obtained with the proposed FTC scheme (4.38) (solid line) and without it (dashed line). The angles are estimated correctly by the observer (4.39). When the actuator fault (4.95) is applied to the TRMS without the proposed FTC scheme, the pitch angle diverges from the desired reference and presents a higher oscillation compared to the case when the FTC scheme is used. On the other hand, although the azimuth angle tracks the reference, an oscillation appears when the FTC scheme is not used. Figure 4.6a shows how the control action

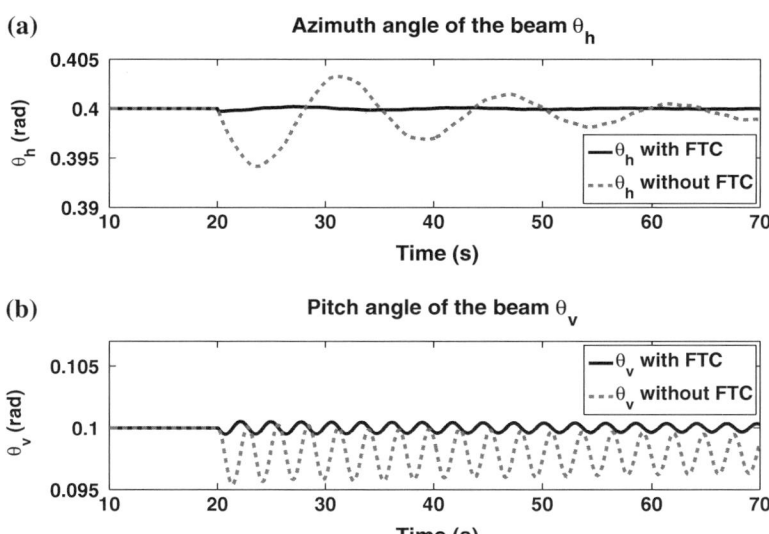

Fig. 4.5 Fault scenario 2: **a** azimuth angle of the beam (*horizontal position*), **b** pitch angle of the beam (*vertical position*)

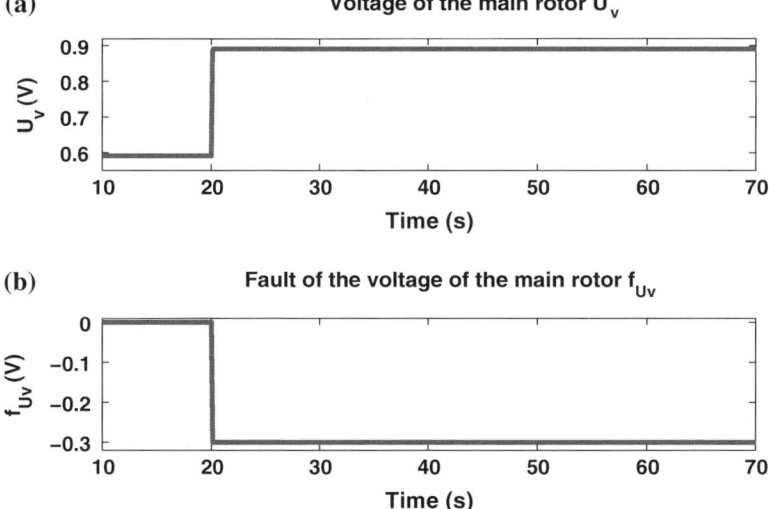

Fig. 4.6 Fault scenario 2: **a** voltage of the main rotor U_v, **b** estimated fault of the voltage of the main rotor U_v

$U_{v,k}$ changes to compensate the actuator fault (4.95). Thus, when the FTC scheme is used, the control action $U_{h,k}$ is not affected by the fault. Figure 4.6b presents the estimated fault provided by (4.36) that corresponds with the fault size introduced in (4.95).

4.4.5 Fault Scenario 3

In this last fault scenario, two faults affecting to the main and tail rotors are applied at the same time:

$$f_h = \begin{cases} 0V, & \text{for } t < 20s, \\ -0.3V, & \text{for } t \geq 20s, \end{cases} \tag{4.96}$$

$$f_v = \begin{cases} 0V, & \text{for } t < 20s, \\ -0.4V, & \text{for } t \geq 20s, \end{cases} \tag{4.97}$$

at the equilibrium point $\theta_{h,ref} = 0.4$ rad and $\theta_{v,ref} = 0.1$ rad.

Figure 4.7a presents the response of the azimuth angle of the beam while Fig. 4.7b shows the pitch angle of the beam that are obtained with the proposed FTC scheme (4.38) (solid line) and without it (dashed line). Notice that the TRMS is stabilized in spite of the actuator faults (4.96) and (4.97). However, when the actuator faults are applied to the TRMS without the proposed FTC scheme, the TRMS cannot follow the reference for the azimuth and pitch angles. Figure 4.8 shows the control action $u_{f,k}$ provided by the FTC scheme (4.38). It can be seen that the control action changes to compensate the actuator faults. Finally, Fig. 4.9 presents the estimated faults provided by (4.36). Notice that both actuator faults were estimated with very high accuracy according to the fault definitions (4.96) and (4.97).

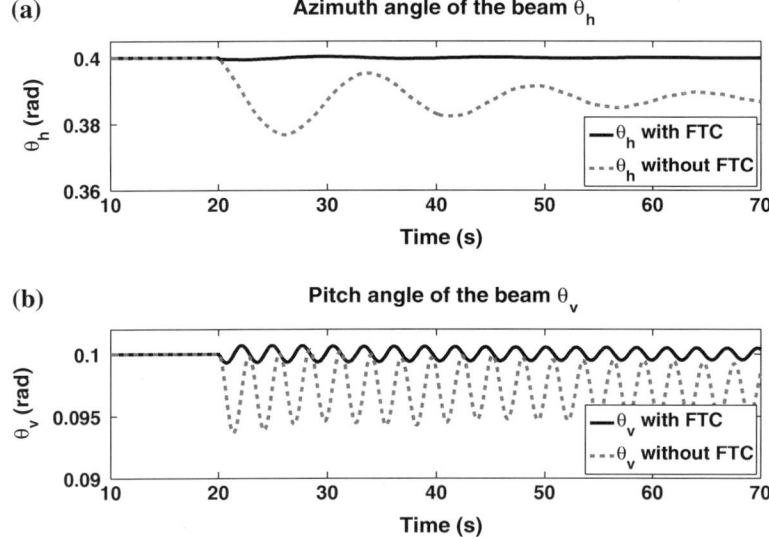

Fig. 4.7 Fault scenario 3: **a** azimuth angle of the beam (*horizontal position*), **b** pitch angle of the beam (*vertical position*)

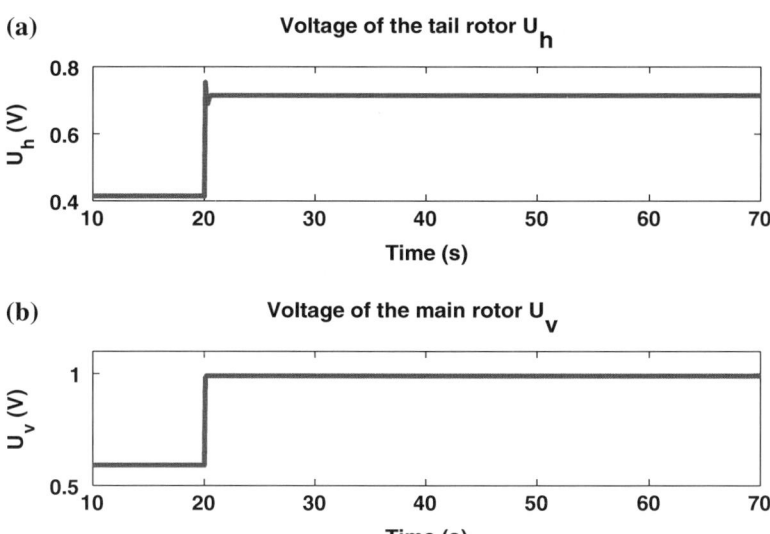

Fig. 4.8 Fault scenario 3: **a** voltage of the tail rotor U_h, **b** voltage of the main rotor U_v

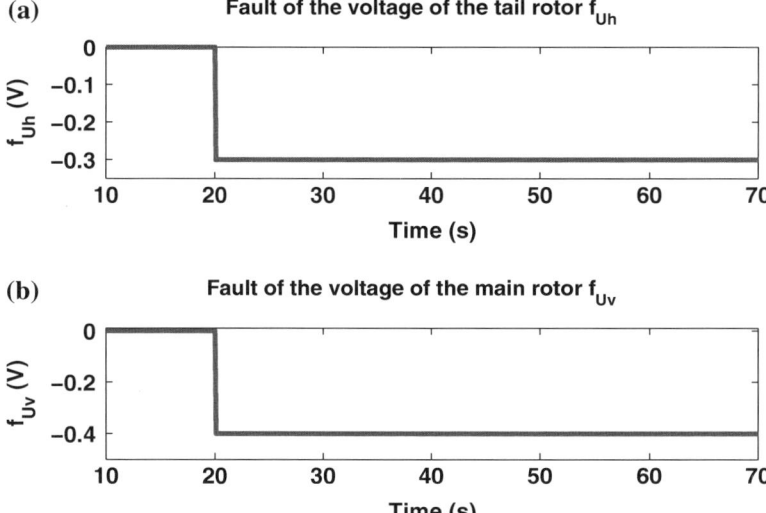

Fig. 4.9 Fault scenario 3: **a** Estimated fault of the voltage of the tail rotor U_h, **b** estimated fault of the voltage of the main rotor U_v

4.5 Concluding Remarks

This chapter proposed a novel FTC scheme that behaves as a virtual actuator, as well as a design procedure that integrates fault identification and fault-tolerant control strategies for a class of non-linear discrete-time systems. In the introductory part of the paper, the proposed fault identification and control strategies were presented for linear systems. Then, these two strategies were extended to the class of non-linear discrete-time systems considered. As a result, non-linear observer-based fault identification and FTC strategies were proposed. It was also shown that for the class of non-linear systems considered the design of the observer and FTC scheme can be done separately. This is a very important contribution of the chapter towards the state-of-the-art in FTC. It is well known that the separation principle may not hold for non-linear systems. Indeed, it is shown that for the proposed scheme the separation principle is valid. Moreover, as was mentioned in the preceding part of the book, in contrast to the approaches presented in the literature the proposed approach takes into account the fault identification error.

Another important contribution of the chapter is that the problems of designing a non-linear observer and FTC scheme can be effectively solved with the use of linear matrix inequalities. This allows using efficient and widely accessible tools. There is no doubt that this result increases the spectrum of possible practical applications of the proposed approach.

The last part of the chapter shows a comprehensive case study regarding the application of the proposed approach to the twin-rotor system manufactured by Feedback Instruments Limited. The presented results show a comparative study of fault-tolerant control and control without fault tolerance. From the presented results, it is evident that the proposed FTC approach is superior over the control strategy without fault tolerance. Indeed, this case study clearly shows the technological applicability of the proposed approach.

The subsequent three chapters extended the proposed approach to the uncertain case where the process and measurement noise/disturbances (including model uncertainty) play a significant role. To tackle such a challenging problem the H_∞ approach is applied (see, e.g., [32] for a general idea). Moreover, a constraint regarding the Lipschitz condition is relaxed by introducing the class of non-linear systems proposed in [33].

References

1. Y. Zhang, J. Jiang, Bibliographical review on reconfigurable fault-tolerant control systems, *IFAC Symposium Fault Detection Supervision and Safety of Technical Processes, SAFE-PROCESS* (Washington, D.C., USA, 2003), pp. 265–276
2. Y. Liang, D. Liaw, T. Lee, Reliable control of nonlinear systems. IEEE Trans. Autom. Control **45**(4), 706–710 (2000)
3. F. Liao, J. Wang, G. Yang, Reliable robust flight tracking control: an LMI approach. IEEE Trans. Control Syst. Technol. **10**(1), 76–89 (2000)

4. Z. Qu, C.M. Ihlefeld, J. Yufang, A. Saengdeejing, Robust fault-tolerant self-recovering control of nonlinear uncertain systems. Automatica **39**(10), 1763–1771 (2003)
5. J. Korbicz, J. Kościelny, Z. Kowalczuk, W. Cholewa (eds.), *Fault Diagnosis. Models, Artificial Intelligence, Applications* (Springer, Berlin, 2004)
6. H. Li, Q. Zhao, Z. Yang, Reliability modeling of fault tolerant control systems. Int. J. Appl. Math. Comput. Sci. **17**(4), 491–504 (2007)
7. M. Witczak, Advances in model-based fault diagnosis with evolutionary algorithms and neural networks. Int. J. Appl. Math. Comput. Sci. **16**(1), 85–99 (2006)
8. M. Witczak, *Modelling and Estimation Strategies for Fault Diagnosis of Non-linear Systems* (Springer, Berlin, 2007)
9. M. Blanke, M. Kinnaert, J. Lunze, M. Staroswiecki, *Diagnosis and Fault-Tolerant Control* (Springer, New York, 2003)
10. R. Iserman, *Fault Diagnosis Applications: Model Based Condition Monitoring, Actuators, Drives, Machinery, Plants, Sensors, and Fault-tolerant Systems* (Springer, Berlin, 2011)
11. M. Mahmoud, J. Jiang, Y. Zhang, *Active Fault Tolerant Control Systems: Stochastic Analysis and Synthesis* (Springer, Berlin, 2003)
12. H. Noura, D. Theilliol, J. Ponsart, A. Chamseddine, *Fault-tolerant Control Systems: Design and Practical Applications* (Springer, Berlin, 2003)
13. G. Ducard, *Fault-tolerant Flight Control and Guidance Systems: Practical Methods for Small Unmanned Aerial Vehicles* (Springer, Berlin, 2009)
14. J. Maciejowski, *Predictive Control with Constraints* (Prentice Hall, New Jersey, 2002)
15. S. Kanev, M. Verhaegen, Reconfigurable robust fault-tolerant control and state estimation, in *15th IFAC World Congress* (Barcelona, Spain, 2002), pp. 1–6
16. J. Lunze, T. Steffen, Control reconfiguration after actuator failures using disturbance decoupling methods. IEEE Trans. Autom. Control **51**(10), 1590–1601 (2006)
17. J.H. Richter, S. Weiland, W.P.M.H. Heemels, J. Lunze, Decoupling-based reconfigurable control of linear systems after actuator faults, in *10th European Control Conference, ECC* (Budapest, Hungary, 2009), pp. 2512–2517
18. J.H. Richter and J. Lunze. H-infinity-based virtual actuator synthesis for optimal trajectory recovery, in *7th IFAC Symposium on Fault Detection, Supervision and Safety of Technical Processes, SAFEPROCESS* (Barcelona, Spain, 2009), pp. 1587–1592
19. L. Dziekan, M. Witczak, J. Korbicz, Active fault-tolerant control design for Takagi-Sugeno fuzzy systems. Bull. Pol. Acad. Sci.: Tech. Sci. **59**(1), 93–102 (2011)
20. S. de Oca, V. Puig, M. Witczak, L. Dziekan, Fault-tolerant control strategy for actuator faults using LPV techniques: Application to a two degree of freedom helicopter. Int. J. Appl. Math. Comput. Sci. **22**(1), 161–171 (2012)
21. M. Witczak, J. Korbicz, A fault-tolerant control strategy for lipschitz non-linear discrete-time systems, in *18th Mediterranean Conference on Control and Automation* (Marrakech, Maroko, 2010), pp. 1079–1084
22. M. Witczak, J. Korbicz, A fault-tolerant control scheme for non-linear discrete-time systems, in *Methods and Models in Automation and Robotics-MMAR 2010: Proceedings of the 15th International Conference* (Miedzyzdroje, Polska, 2010), pp. 302–307
23. M. Witczak, V. Puig, S. de Oca, A fault-tolerant control strategy for non-linear discrete-time systems: application to the twin-rotor system. Int. J. Control. **86**(10), 1788–1799 (2013)
24. M. Witczak, L. Dziekan, V. Puig, J. Korbicz, An integrated design strategy for fault identification and fault-tolerant control for Takagi-Sugeno fuzzy systems, in *Workshop on Advanced Control and Diagnosis, ACD* (Grenoble, France, 2007), pp. 1–6
25. B. Jiang, F.N. Chowdhury, Fault estimation and accommodation for linear MIMO discrete-time systems. IEEE Trans. Control Syst. Technol. **13**(1), 493–499 (2005)
26. S. Hui, S.H. Zak, Observer design for systems with unknown input. Int. J. Appl. Math. Comput. Sci. **15**(4), 431–446 (2005)
27. D. Nesic, A. Loria, On uniform asymptotic stability of time-varying parameterized discrete-time cascades. IEEE Trans. Autom. Control **49**(6), 875–887 (2004)

28. T.C. Lee, Z.P. Jiang, On uniform asymptotic stability of time-varying parameterized discrete-time cascades. IEEE Trans. Autom. Control **51**(10), 1644–1660 (2006)
29. A. Zemouche, M. Boutayeb, Observer design for Lipschitz non-linear systems: the discrete time case. IEEE Trans. Circ. Syst. - II:Express Briefs **53**(8), 777–781 (2006)
30. D. Simon, T.L. Chia, Kalman filtering with state equality constraints. IEEE Trans. Aerosp. Electron. Syst. **38**(1), 128–136 (2002)
31. A. Rahideh, M.H. Shaheed, Mathematical dynamic modelling of a twin-rotor multiple input-multiple output system, in *Institution of Mechanical Engineers, Part I: Journal of Systems and Control Engineering* **227**, 89–101 (2007)
32. H. Li, M. Fu, A linear matrix inequality approach to robust h_∞ filtering. IEEE Trans. Signal Process. **45**(9), 2338–2350 (1997)
33. D.M. Stipanovic, D.D. Siljak, Robust stability and stabilization of discrete-time non-linear: the lmi approach. Int. J. Control **74**(5), 873–879 (2001)

Chapter 5
Robust $\mathcal{H}_\infty$-Based Approaches

5.1 Towards Robust Fault-Tolerant Control

The main objective of the present section is to extend the FTC scheme presented in the preceding chapter by introducing the following improvements:

- the class of non-linear system is extended towards the one presented in [1];
- the $\mathcal{H}_\infty$ approach is applied to achieve robustness with respect to the exogenous disturbance present in the state equation;
- the effect of the one-step fault prediction error discussed in the previous chapter is minimised with the $\mathcal{H}_\infty$ approach.

The proposed approach can be perceived as a combination of linear-system strategies [2] and [3] for a class of non-linear systems [1, 4]. It is designed in such a way that the prescribed disturbance attenuation level is achieved with respect to the fault estimation error while guaranteeing the convergence of the observer. The same goal is attained in the control framework where a prescribed disturbance attenuation level is achieved with respect to the tracking error. The resulting design procedure boils down to solving a set of linear matrix inequalities, which can be easily realised with modern computational packages.

The section starts with the introduction of a new class of non-linear systems. Subsequently, the robust fault estimation approach is proposed. Based on the achieved results, robust FTC is proposed.

5.1.1 Preliminaries

Let us consider a non-linear discrete-time system (possibly faulty):

$$x_{f,k+1} = Ax_{f,k} + Bu_{f,k} + g(x_{f,k}) + Lf_k + Ww_k, \qquad (5.1)$$

$$y_{f,k+1} = Cx_{f,k+1}, \qquad (5.2)$$

M. Witczak, *Fault Diagnosis and Fault-Tolerant Control Strategies for Non-Linear Systems*, 143
Lecture Notes in Electrical Engineering 266, DOI: 10.1007/978-3-319-03014-2_5,
© Springer International Publishing Switzerland 2014

where $x_{f,k} \in \mathbb{X} \subset \mathbb{R}^n$ is the state, $u_{f,k} \in \mathbb{R}^r$ stands for the input, $y_{f,k} \in \mathbb{R}^m$ denotes the output, $f_k \in \mathbb{R}^s$ stands for the fault, $w_k \in l_2^n$ is a an exogenous disturbance vector and $W \in \mathbb{R}^{n \times n}$ stands for its distribution matrix, while:

$$l_2 = \left\{ \mathbf{w} \in \mathbb{R}^n \mid \|\mathbf{w}\|_{l_2} < +\infty \right\}, \tag{5.3}$$

$$\|\mathbf{w}\|_{l_2} = \left(\sum_{k=0}^{\infty} \|w_k\|^2 \right)^{\frac{1}{2}}. \tag{5.4}$$

Moreover, $g(x)$ is a non-linear function satisfying [1, 4]

$$(g(x_1) - g(x_2))^T (g(x_1) - g(x_2))$$
$$\leq (x_1 - x_2)^T M^T M (x_1 - x_2), \ \forall x_1, x_2 \in \mathbb{X}, \tag{5.5}$$

where $M \in \mathbb{R}^{n \times n}$. In the subsequent part of this section it will be shown how to deal with such a system representation.

Using the Differential Mean Value Theorem (DMVT) [5], it can be shown that

$$g(a) - g(b) = M_x(a - b), \tag{5.6}$$

with

$$M_x = \begin{bmatrix} \dfrac{\partial g_1}{\partial x}(c_1) \\ \vdots \\ \dfrac{\partial g_n}{\partial x}(c_n) \end{bmatrix}, \tag{5.7}$$

where $c_1, \ldots, c_n \in \mathrm{Co}(a, b)$, $c_i \neq a$, $c_i \neq b$, $i = 1, \ldots, n$. Assuming that

$$\bar{a}_{i,j} \geq \frac{\partial g_i}{\partial x_j} \geq \underline{a}_{i,j}, \quad i = 1, \ldots, n, \quad j = 1, \ldots, n, \tag{5.8}$$

it is clear that there exists a matrix $M \in \mathbb{M}$:

$$\mathbb{M} = \left\{ M \in \mathbb{R}^{n \times n} \mid \bar{a}_{i,j} \geq m_{i,j} \geq \underline{a}_{i,j}, \ i, j = 1, \ldots, n, \right\}, \tag{5.9}$$

for which $M_x^T M_x \preceq M^T M$. In order to find the upper bound $M^T M$, the following evident inequality is used: $M^T M \preceq \lambda_{\max}(M^T M) I_n$. Thus, the problem is

$$M^* = \arg \max_{M \in \mathbb{M}} \lambda_{\max}(M^T M). \tag{5.10}$$

Taking into account the fact that $\lambda_{\max}(M^T M) = \|M^T M\|_2 = \|M\|_2^2$, the optimisation problem (5.10) can be replaced by

$$M^* = \arg \max_{M \in \mathbb{M}} \|M\|, \qquad (5.11)$$

which can be perceived as a worst case norm analysis task. This can be easily solved, e.g., with MATLAB (cf. the **wcnorm** function).

5.1.2 Fault Estimation Approach

Following [2, 6], let us assume that the system is observable and the following rank condition is satisfied:

$$\text{rank}(CL) = \text{rank}(L) = s. \qquad (5.12)$$

Under the assumption (5.12), it is possible to calculate $H = (CL)^+ = \left[(CL)^T CL\right]^{-1} (CL)^T$. By multiplying (5.2) by H and then substituting (5.1), it can be shown that

$$f_k = H(y_{f,k+1} - CAx_{f,k} - CBu_{f,k} - Cg(x_{f,k}) - CWw_k). \qquad (5.13)$$

Finally, by substituting (5.13) into (5.1) it can be shown that

$$x_{f,k+1} = \bar{A}x_{f,k} + \bar{B}u_{f,k} + Gg(x_{f,k}) + \bar{L}y_{f,k+1} + \bar{W}w_k, \qquad (5.14)$$

where $G = (I_n - LHC)$, $\bar{A} = GA$, $\bar{B} = GB$, $\bar{L} = LH$, $\bar{W} = GW$. In order to estimate (5.13), i.e., to obtain $\hat{f}_k$, it is necessary to estimate the state of the system, i.e., to obtain $\hat{x}_{f,k}$. Consequently, the fault estimate is given as follows:

$$\hat{f}_k = H\left(y_{f,k+1} - CA\hat{x}_{f,k} - CBu_{f,k} - Cg(\hat{x}_{f,k})\right). \qquad (5.15)$$

The corresponding observer structure is

$$\hat{x}_{f,k+1} = \bar{A}\hat{x}_{f,k} + \bar{B}u_{f,k} + Gg(\hat{x}_{f,k}) + \bar{L}y_{f,k+1} + K_3(y_{f,k} - C\hat{x}_{f,k}), \qquad (5.16)$$

while the state estimation error is given by

$$\begin{aligned} e_{f,k+1} &= \left(\bar{A} - K_3 C\right) e_{f,k} + Gs_k + \bar{W}w_k \\ &= A_1 e_{f,k} + Gs_k + \bar{W}w_k, \end{aligned} \qquad (5.17)$$

where

$$s_k = g(x_{f,k}) - g(\hat{x}_{f,k}). \qquad (5.18)$$

Similarly, the fault estimation error $\varepsilon_{f,k}$ can be defined as

$$\varepsilon_{f,k} = f_k - \hat{f}_k = -HC\left(Ae_{f,k} + s_k + Ww_k\right). \qquad (5.19)$$

The objective of further deliberations is to design the observer (5.16) in such a way that the state estimation error $e_{f,k}$ is asymptotically convergent and the following upper bound is guaranteed:

$$\|\varepsilon_f\|_{l_2} \leq \mu \|\mathbf{w}\|_{l_2}, \tag{5.20}$$

where $\mu > 0$ is a prescribed disturbance attenuation level. Thus, contrary to the approaches presented in the literature, μ should be achieved with respect to the fault estimation error but not the state estimation error.

The following theorem constitutes the main result of this section.

Theorem 5.1 *For a prescribed disturbance attenuation level $\mu > 0$ for the fault estimation error (5.19), the $\mathcal{H}_\infty$ observer design problem for the system (5.1) and (5.2) and the observer (5.16) is solvable if there exist $\alpha > 0$, $\beta > 0$, $P \succ 0$, $Q \succ 0$, N such that the following LMI is satisfied:*

$$\begin{bmatrix} -P + A^T \bar{H} A + \beta M^T M & A^T \bar{H} & A^T \bar{H} W & 0 & \bar{A}^T P - C^T N^T \\ \bar{H} A & \bar{H} - Q - \beta I & \bar{H} W & 0 & G^T P \\ W^T \bar{H} A & W^T \bar{H} & W^T \bar{H} W - \mu^2 I & 0 & W^T G^T P \\ 0 & 0 & 0 & Q - \alpha I & 0 \\ P\bar{A} - NC & PG & PGW & 0 & -\frac{1}{2}P \end{bmatrix} \prec 0, \tag{5.21}$$

along with (5.35).

proof The problem of $\mathcal{H}_\infty$ observer design [4, 7] is to determine the gain matrix K_3 such that

$$\lim_{k\to\infty} e_{f,k} = 0 \quad \text{for} \quad \mathbf{w}_k = 0, \tag{5.22}$$

$$\|\varepsilon_f\|_{l_2} \leq \mu \|\mathbf{w}\|_{l_2} \quad \text{for} \quad \mathbf{w}_k \neq 0, \; e_{f,0} = 0. \tag{5.23}$$

In order to settle the above problem it is sufficient to find a Lyapunov function V_k such that

$$\Delta V_k + \varepsilon_{f,k}^T \varepsilon_{f,k} - \mu^2 \mathbf{w}_k^T \mathbf{w}_k < 0, \; k = 0, \dots \infty, \tag{5.24}$$

where $\Delta V_k = V_{k+1} - V_k$. Indeed, if $\mathbf{w}_k = 0$ then (5.24) boils down to

$$\Delta V_k + \varepsilon_{f,k}^T \varepsilon_{f,k} < 0, \; k = 0, \dots \infty, \tag{5.25}$$

and hence $\Delta V_k < 0$, which leads to (5.22). If $\mathbf{w}_k \neq 0$, then (5.24) yields

$$J = \sum_{k=0}^{\infty} \left(\Delta V_k + \varepsilon_{f,k}^T \varepsilon_{f,k} - \mu^2 \mathbf{w}_k^T \mathbf{w}_k \right) < 0, \tag{5.26}$$

which can be written as

$$J = -V_0 + \sum_{k=0}^{\infty} \varepsilon_{f,k}^T \varepsilon_{f,k} - \sum_{k=0}^{\infty} \mu^2 \boldsymbol{w}_k^T \boldsymbol{w}_k < 0. \tag{5.27}$$

Knowing that $V_0 = 0$ for $\boldsymbol{e}_{f,0} = 0$, (5.27) leads to (5.23).

Since the general framework for designing the robust observer is given, the following form of the Lyapunov function is proposed [5]:

$$V_k = \boldsymbol{e}_{f,k}^T \boldsymbol{P} \boldsymbol{e}_{f,k} + \boldsymbol{s}_k^T \boldsymbol{Q} \boldsymbol{s}_k, \tag{5.28}$$

where $\boldsymbol{P} \succ \boldsymbol{0}$ and $\boldsymbol{Q} \succ \boldsymbol{0}$. Consequently,

$$
\begin{aligned}
\Delta V_k + \varepsilon_{f,k}^T \varepsilon_{f,k} &- \mu^2 \boldsymbol{w}_k^T \boldsymbol{w}_k \\
&= \boldsymbol{e}_{f,k}^T \left(\boldsymbol{A}_1^T \boldsymbol{P} \boldsymbol{A}_1 - \boldsymbol{P} - \boldsymbol{A}^T \bar{\boldsymbol{H}} \boldsymbol{A} \right) \boldsymbol{e}_{f,k} \\
&+ \boldsymbol{e}_{f,k}^T \left(\boldsymbol{A}_1^T \boldsymbol{P} \boldsymbol{G} + \boldsymbol{A}^T \bar{\boldsymbol{H}} \right) \boldsymbol{s}_k \\
&+ \boldsymbol{e}_{f,k}^T \left(\boldsymbol{A}_1^T \boldsymbol{P} \boldsymbol{G} \boldsymbol{W} + \boldsymbol{A}^T \bar{\boldsymbol{H}} \boldsymbol{W} \right) \boldsymbol{w}_k \\
&+ \boldsymbol{s}_k^T \left(\boldsymbol{G}^T \boldsymbol{P} \boldsymbol{A}_1 - \bar{\boldsymbol{H}} \boldsymbol{A} \right) \boldsymbol{e}_{f,k} \\
&+ \boldsymbol{s}_k^T \left(\boldsymbol{G}^T \boldsymbol{P} \boldsymbol{G} + \bar{\boldsymbol{H}} - \boldsymbol{Q} \right) \boldsymbol{s}_k \\
&+ \boldsymbol{s}_k^T \left(\boldsymbol{G}^T \boldsymbol{P} \boldsymbol{G} \boldsymbol{W} + \bar{\boldsymbol{H}} \boldsymbol{W} \right) \boldsymbol{w}_k \\
&+ \boldsymbol{w}_k^T \left(\boldsymbol{W}^T \boldsymbol{G}^T \boldsymbol{P} \boldsymbol{A}_1 + \boldsymbol{W}^T \bar{\boldsymbol{H}} \boldsymbol{A} \right) \boldsymbol{e}_{f,k} \\
&+ \boldsymbol{w}_k^T \left(\boldsymbol{W}^T \boldsymbol{G}^T \boldsymbol{P} \boldsymbol{G} + \boldsymbol{W}^T \bar{\boldsymbol{H}} \right) \boldsymbol{s}_k \\
&+ \boldsymbol{w}_k^T \left(\boldsymbol{W}^T \boldsymbol{G}^T \boldsymbol{P} \boldsymbol{G} \boldsymbol{W} + \boldsymbol{W}^T \bar{\boldsymbol{H}} \boldsymbol{W} - \mu^2 \boldsymbol{I}_n \right) \boldsymbol{w}_k \\
&+ \boldsymbol{s}_{k+1}^T \boldsymbol{Q} \boldsymbol{s}_{k+1} < 0, \tag{5.29}
\end{aligned}
$$

with $\bar{\boldsymbol{H}} = \boldsymbol{C}^T \boldsymbol{H}^T \boldsymbol{H} \boldsymbol{C}$. By defining

$$\boldsymbol{v}_k = \left[\boldsymbol{e}_{f,k}^T, \, \boldsymbol{s}_k^T, \, \boldsymbol{w}_k^T, \, \boldsymbol{s}_{k+1}^T \right]^T, \tag{5.30}$$

the inequality (5.29) becomes

$$\Delta V_k + \varepsilon_{f,k}^T \varepsilon_{f,k} - \mu^2 \boldsymbol{w}_k^T \boldsymbol{w}_k = \boldsymbol{v}_k^T \boldsymbol{M}_V \boldsymbol{v}_k < 0, \tag{5.31}$$

where $\boldsymbol{M}_V$ is given by (5.32):

$$M_V = \begin{bmatrix} A_1^T P A_1 - P + A^T \bar{H} A & A_1^T P G + A_1^T \bar{H} & A_1^T P G W + A_1^T \bar{H} W & 0 \\ G^T P A_1 + \bar{H} A & G^T P G + \bar{H} - Q & G^T P G W + \bar{H} W & 0 \\ W^T G^T P A_1 + W^T \bar{H} A & W^T G^T P G + W^T \bar{H} & W^T G^T P G W + W^T \bar{H} W - \mu^2 I & 0 \\ 0 & 0 & 0 & Q \end{bmatrix}, \quad (5.32)$$

Additionally, from (5.5),

$$\beta e_{f,k}^T M^T M e_{f,k} - \beta s_k^T s_k \geq 0, \quad \beta > 0, \tag{5.33}$$

which is equivalent to

$$v_k^T \begin{bmatrix} \beta M^T M & 0 & 0 & 0 \\ 0 & -\beta I & 0 & 0 \\ 0 & 0 & 0 & 0 \\ 0 & 0 & 0 & 0 \end{bmatrix} v_k \geq 0. \tag{5.34}$$

Similarly, from (5.5) and by assuming that

$$P \succ \alpha M^T M, \quad \alpha > 0, \tag{5.35}$$

it can be shown that

$$\alpha e_{f,k+1}^T M^T M e_{f,k+1} - \alpha s_{k+1}^T s_{k+1}$$
$$< e_{f,k+1}^T P e_{f,k+1} - \alpha s_{k+1}^T s_{k+1} \geq 0, \tag{5.36}$$

which is equivalent to

$$v_k^T M_Y v_k \geq 0, \tag{5.37}$$

with

$$M_Y = \begin{bmatrix} A_1^T P A_1 & A_1^T P G & A_1^T P G W & 0 \\ G^T P A_1 & G^T P G & G^T P G W & 0 \\ W^T G^T P A_1 & W^T G^T P G & W^T G^T P G W & 0 \\ 0 & 0 & 0 & -\alpha I \end{bmatrix}. \tag{5.38}$$

Finally, using (5.31) along with (5.34) and (5.37), the convergence condition of the observer becomes

$$
\begin{bmatrix}
2A_1^T P A_1 - P + A^T \bar{H} A + \beta M^T M & 2A_1^T P G + A_1^T \bar{H} \\
2G^T P A_1 + \bar{H} A & 2G^T P G + \bar{H} - Q - \beta I \\
2W^T G^T P A_1 + W^T \bar{H} A & 2W^T G^T P G + W^T \bar{H} \\
\mathbf{0} & \mathbf{0}
\end{bmatrix}
$$

$$
\qquad\qquad
\begin{bmatrix}
2A_1^T P G W + A_1^T \bar{H} W & \mathbf{0} \\
2G^T P G W + \bar{H} W & \mathbf{0} \\
2W^T G^T P G W + W^T \bar{H} W - \mu^2 I & \mathbf{0} \\
\mathbf{0} & Q - \alpha I
\end{bmatrix} \prec \mathbf{0}.
\tag{5.39}
$$

Moreover, by applying the Schur complements, (5.39) can be transformed into

$$
\begin{bmatrix}
-P + A^T \bar{H} A + \beta M^T M & A^T \bar{H} & A^T \bar{H} W & 0 & A_1^T \\
\bar{H} A & \bar{H} - Q - \beta I & \bar{H} W & 0 & G^T \\
W^T \bar{H} A & W^T \bar{H} & W^T \bar{H} W - \mu^2 I & 0 & W^T G^T \\
0 & 0 & 0 & Q - \alpha I & 0 \\
A_1 & G & GW & 0 & -\frac{1}{2} P^{-1}
\end{bmatrix} \prec 0.
\tag{5.40}
$$

Multiplying (5.40) from both sites by

$$
\mathrm{diag}(I, I, I, P)
\tag{5.41}
$$

and then substituting

$$
P A_1 = P \bar{A} - P K_3 C = P \bar{A} - N C
\tag{5.42}
$$

yield (5.21), which completes the proof.

Note that (5.21) is a usual LMI, which can be easily solved, e.g., with MATLAB. Thus, the final design procedure is as follows: given a prescribed disturbance attenuation level μ, obtain $\alpha > 0$, $\beta > 0$, $P \succ 0$, $Q \succ 0$, N by solving (5.35) and (5.21). Finally, the gain matrix of (5.16) is

$$
K_3 = P^{-1} N.
\tag{5.43}
$$

It can be also observed that the observer design problem can be treated as an minimisation task, i.e.,

$$
\mu^* = \min_{\mu > 0, \alpha > 0, \beta > 0, P \succ 0, Q \succ 0, N} \mu,
\tag{5.44}
$$

under (5.35) and (5.21).

5.1.3 Integrated FTC Design

Similarly as in Chap. 4, let us consider a reference model given by

$$x_{k+1} = Ax_k + Bu_k + g(x_k), \tag{5.45}$$
$$y_{k+1} = Cx_{k+1}, \tag{5.46}$$

and $L = B$ in (5.1) and (5.2).

The objective is to design the control strategy $u_{f,k}$ for (5.1) and (5.2) such that the tracking error

$$e_k = x_k - x_{f,k}, \tag{5.47}$$

will be asymptotically convergent, guaranteeing the prescribed disturbance attenuation level. To achieve this goal, the following control strategy is proposed:

$$u_{f,k} = -\hat{f}_{k-1} + K_1(x_k - \hat{x}_{f,k}) + K_2\gamma_k + u_k. \tag{5.48}$$

Taking into account the problems with one-step fault prediction (cf. Chap. 4), the following assumption is imposed:

$$\hat{f}_k = \hat{f}_{k-1} + \bar{v}_k, \quad \bar{v}_k \in l_2. \tag{5.49}$$

Bearing in mind that all faults present in real systems have a finite value, such an assumption is fully justified. Thus, for convergence analysis, the following form of the FTC control is used:

$$u_{f,k} = -\hat{f}_k + \bar{v}_k + K_1(x_k - \hat{x}_{f,k}) + K_2\gamma_k + u_k, \tag{5.50}$$

Using a similar approach as in Chap. 4 and setting $K_2 = HC$, the tracking error becomes

$$e_{k+1} = A_1 e_k + (BHCA - BK_1)e_{f,k} + G\omega_k + \tilde{W}\bar{w}_k, \tag{5.51}$$

with $K_2 = HC$, $A_1 = A - BK_1$ and $H = (CB)^+$, where

$$\tilde{W} = [B, \ [BHC - I]W], \quad \bar{w}_k = \begin{bmatrix} \bar{v}_k \\ w_k \end{bmatrix}. \tag{5.52}$$

Using the same arguments as in Sect. 4.2, the convergence analysis can be relaxed to the following form of the tracking error:

$$e_{k+1} = A_1 e_k + G\omega_k + \tilde{W}\bar{w}_k. \tag{5.53}$$

The following theorem constitutes the main result of the present section.

Theorem 5.2 *For a prescribed disturbance attenuation level $\mu > 0$ for the tracking error (5.53), the $\mathcal{H}_\infty$ controller design problem (5.50) for the system (5.1) and (5.2) is solvable if there exist $\alpha > 0$, $\beta > 0$, $P \succ 0$, $Q \succ 0$, N such that the following LMI is satisfied:*

$$\begin{bmatrix} I - P + \beta M^T M & 0 & 0 & 0 & A_1 P \\ 0 & -Q - \beta I & 0 & 0 & G P \\ 0 & 0 & -\mu^2 I & 0 & \tilde{W} P \\ 0 & 0 & 0 & Q - \alpha I & 0 \\ P A_1^T & P G^T & P \tilde{W}^T & 0 & -\frac{1}{2} P \end{bmatrix} \prec 0, \qquad (5.54)$$

along with (5.35), *where*

$$A_1 P = AP - BK_1 P = AP - BN. \qquad (5.55)$$

proof   By defining the Lyapunov function

$$V_k = e_k^T P e_k + \omega_k^T Q \omega_k \qquad (5.56)$$

and

$$\Delta V_k + e_k^T e_k - \mu^2 \bar{w}_k^T \bar{w}_k < 0, \qquad (5.57)$$

as well as using the same line of reasoning as in Theorem 5.1, it is possible derive (5.54), which completes the proof.

Thus, the final design procedure is as follows: given a prescribed disturbance attenuation level μ, obtain $\alpha > 0$, $\beta > 0$, $P \succ 0$, $Q \succ 0$, N by solving (5.35) and (5.54). Finally, the gain matrix of the FTC controller is

$$K_1 = N P^{-1}. \qquad (5.58)$$

5.1.4 Illustrative Example: Fault Estimation

Let us consider a non-linear system

$$x_{f,k+1} = A x_{f,k} + B u_{f,k} + g(x_{f,k}) + L f_k + W w_k, \qquad (5.59)$$
$$y_{f,k+1} = C x_{f,k+1}, \qquad (5.60)$$

with

$$A = \begin{bmatrix} 0.137 & 0.199 & 0.284 \\ 0.0118 & 0.299 & 0.47 \\ 0.894 & 0.661 & 0.065 \end{bmatrix}, \quad B = \begin{bmatrix} 0.25 \\ 0.6 \\ 0.1 \end{bmatrix},$$

$$C = \begin{bmatrix} 1 & 0 & 0 \\ 0 & 1 & 0 \end{bmatrix}, \quad L = \begin{bmatrix} 0 \\ 1 \\ 0 \end{bmatrix}, \quad W = \begin{bmatrix} 0.5 & 0 & 0 \\ 0 & 0.4 & 0 \\ 0 & 0 & 0.3 \end{bmatrix},$$

and

$$g(x_{f,k}) = \left[\frac{0.6\cos(12x_{1,k})}{x_{2,k}^2 + 10},\ 0,\ -0.33\sin(x_{3,k}) \right]^T. \tag{5.61}$$

Since the system is given, it is straightforward to calculate

$$M_x^T M_x = \mathrm{diag}\left(0, \left(\frac{-7.2\sin(12x_{2,k})}{x_{2,k}^2 + 10} - \frac{1.2\cos(12x_{2,k})x_{2,k}}{(x_{2,k}^2 + 10)^2} \right)^2, 0.1089\cos(x_{3,k})^2 \right). \tag{5.62}$$

Finally, the application of the proposed procedure leads to

$$M^T M = \begin{bmatrix} 0 & 0 & 0 \\ 0 & 0.517 & 0 \\ 0 & 0 & 0.1089 \end{bmatrix}. \tag{5.63}$$

As a result of solving the problem (5.44), the following couple was obtained:

$$\mu^* = 0.499, \quad K_3 = \begin{bmatrix} 0.4885 & 2.0619 \\ 0 & 0 \\ 0.9724 & 1.0753 \end{bmatrix}. \tag{5.64}$$

Let the initial condition for the system and the observer be

$$x_{f,0} = [3, 2, 1]^T, \quad \hat{x}_{f,0} = 0, \tag{5.65}$$

while the input and the exogenous disturbance are

$$u_{f,k} = \sin(0.002\pi k), \quad w_k \sim \mathcal{N}(0, 0.1^2 I). \tag{5.66}$$

Moreover, let us consider the following fault scenario:

$$f_k = \begin{cases} 1, & \text{for } 300 \geq k \geq 200, \\ 0, & \text{otherwise.} \end{cases} \tag{5.67}$$

First, let us consider the case when $\hat{x}_{f,0} = x_{f,0}$ ($e_{f,0} = 0$). Figure 5.1 clearly indicates that the condition (5.23) is satisfied, which means that an attenuation level $\mu^* = 0.499$ is achieved.

Now let us assume that $w_k = 0$ and $\hat{x}_{f,0} \neq x_{f,0}$. Figure 5.2 clearly shows that (5.22) is satisfied as well.

Finally, Fig. 5.3 shows the fault and its estimate for the nominal case ($\hat{x}_{f,0} \neq x_{f,0}$ and $w_k \neq 0$).

From these results, it can be observed that the proposed tool can be efficiently applied for solving robust $\mathcal{H}_\infty$-based fault identification of non-linear discrete time systems.

Fig. 5.1 Evolution of ΔV_k $+ \varepsilon_{f,k}^T \varepsilon_{f,k} - \mu^2 \boldsymbol{w}_k^T \boldsymbol{w}_k$

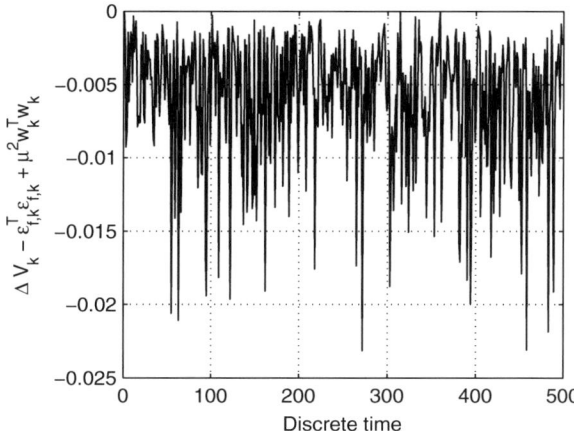

Fig. 5.2 Evolution of $\|\boldsymbol{e}_{f,k}\|$ (for $k = 0, \ldots, 20$)

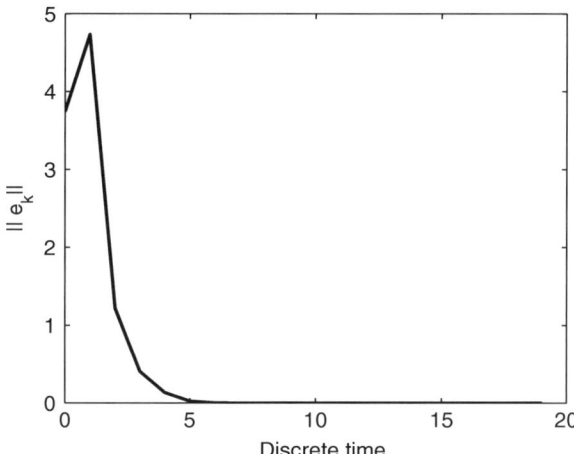

5.2 Complete Robust Design of Fault-Tolerant Control

The present section extends the proposed results into the case when uncertainties are present both in the state and the output equation. The structure of the section is similar to that of the preceding one. It starts with the problem of fault estimation, and then an integrated FTC control strategy is proposed. The final part of the section contains an illustrative example, which shows the performance of the selected methods.

Fig. 5.3 Fault and its estimate

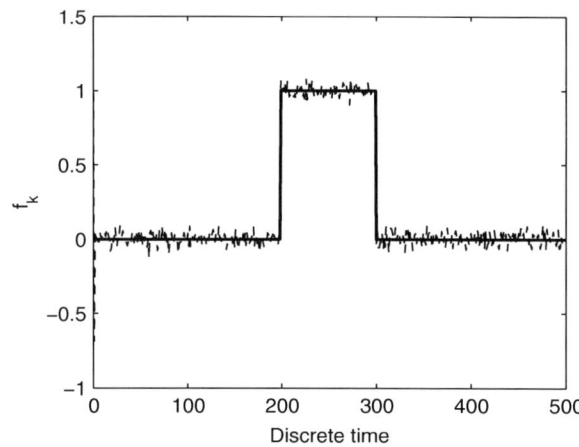

5.2.1 *Preliminaries*

Let us consider a non-linear discrete-time system:

$$x_{f,k+1} = Ax_{f,k} + Bu_{f,k} + g(x_{f,k}) + Lf_k + W_1w_k, \qquad (5.68)$$
$$y_{f,k+1} = Cx_{f,k+1} + W_2w_{k+1}, \qquad (5.69)$$

where $x_{f,k} \in \mathbb{X} \subset \mathbb{R}^n$ is the state, $u_{f,k} \in \mathbb{R}^r$ stands for the input, $y_{f,k} \in \mathbb{R}^m$ denotes the output, $f_k \in \mathbb{R}^s$ stands for the fault, $w_k \in l_2$ is a an exogenous disturbance vector and $W_1 \in \mathbb{R}^{n \times n}$, $W_2 \in \mathbb{R}^{m \times n}$ stand for its distribution matrices, while

$$l_2 = \left\{ \mathbf{w} \in \mathbb{R}^n | \ \|\mathbf{w}\|_{l_2} < +\infty \right\}, \ \|\mathbf{w}\|_{l_2} = \left(\sum_{k=0}^{\infty} \|w_k\|^2 \right)^{\frac{1}{2}}. \qquad (5.70)$$

5.2.2 *Fault Estimation Strategy*

Following [2, 6], let us assume that the system is observable and the following rank condition is satisfied:

$$\mathrm{rank}(CL) = \mathrm{rank}(L) = s. \qquad (5.71)$$

Under the assumption (5.71), it is possible to calculate $H = (CL)^+ = \left[(CL)^T CL \right]^{-1} (CL)^T$. By multiplying (5.69) by H and then substituting (5.68) it can be shown that:

$$f_k = H(y_{f,k+1} - CAx_{f,k} - CBu_{f,k} - Cg(x_{f,k})$$
$$- CW_1w_k - W_2w_{k+1}). \tag{5.72}$$

Finally, by substituting (5.72) into (5.68) it can be shown that:

$$x_{f,k+1} = \bar{A}x_{f,k} + \bar{B}u_{f,k} + Gg(x_{f,k}) + \bar{L}y_{f,k+1} + GW_1w_k$$
$$- \bar{L}W_2w_{k+1}, \tag{5.73}$$

where $G = (I_n - LHC)$, $\bar{A} = GA$, $\bar{B} = GB$, $\bar{L} = LH$. In order to estimate (5.72), i.e., to obtain $\hat{f}_k$, it is necessary to estimate the state of the system, i.e., to obtain $\hat{x}_{f,k}$. Consequently, the fault estimate is given as follows:

$$\hat{f}_k = H\left(y_{f,k+1} - CA\hat{x}_{f,k} - CBu_{f,k} - Cg(\hat{x}_{f,k})\right). \tag{5.74}$$

The corresponding observer structure is

$$\hat{x}_{f,k+1} = \bar{A}\hat{x}_{f,k} + \bar{B}u_{f,k} + Gg(\hat{x}_{f,k}) + \bar{L}y_{f,k+1} + K_3(y_{f,k} - C\hat{x}_{f,k}), \tag{5.75}$$

while the state estimation error is given by

$$e_{f,k+1} = \left(\bar{A} - K_3C\right)e_{f,k} + Gs_k + (GW_1 - K_3W_2)w_k - \bar{L}W_2w_{k+1}$$
$$= A_1e_{f,k} + Gs_k + \bar{W}_1w_k + \bar{W}_2w_{k+1}, \tag{5.76}$$

where

$$s_k = g(x_{f,k}) - g(\hat{x}_{f,k}). \tag{5.77}$$

Similarly, the fault estimation error $\varepsilon_{f,k}$ can be defined as

$$\varepsilon_{f,k} = f_k - \hat{f}_k = -HC\left(Ae_{f,k} + s_k + W_1w_k\right) - HW_2w_{k+1}. \tag{5.78}$$

The objective of further deliberations is to design the observer (5.75) in such a way that the state estimation error $e_{f,k}$ is asymptotically convergent and the following upper bound is guaranteed:

$$\|\varepsilon_f\|_{l_2} \leq \xi\|w\|_{l_2} \tag{5.79}$$

where $\xi > 0$ is a prescribed disturbance attenuation level. Thus, contrary to the approaches presented in the literature, μ should be achieved with respect to the fault estimation error but not the state estimation error.

The following theorem constitutes the main result of this section.

Theorem 5.3 *For a prescribed disturbance attenuation level $\mu > 0$ for the fault estimation error (5.78), the $\mathcal{H}_\infty$ observer design problem for the system (5.68) and (5.69) and the observer (5.75) is solvable if there exist $\alpha > 0$, $\beta > 0$, $P \succ 0$, $Q \succ 0$, N such that the following LMI is satisfied:*

$$
\begin{bmatrix}
-P + A^T H_1 A + \beta M^T M & A^T H_1 & A^T H_1 W_1 \\
H_1 A & H_1 - Q - \beta I & H_1 W_1 \\
W_1^T H_1 A & W_1^T H_1 & W_1^T H_1 W_1 - \mu^2 I \\
H_2^T A & H_2^T & H_2^T W_1 \\
0 & 0 & 0 \\
P\bar{A} - NC & PG & PGW_1 - NW_2
\end{bmatrix}
$$

$$
\begin{matrix}
A^T H_2 & 0 & \bar{A}^T P - C^T N^T \\
H_2 & 0 & G^T P \\
W_1^T H_2 & 0 & W_1^T G^T P - W_2^T N^T \\
W_2^T H^T H W_2 - \mu^2 I & 0 & \bar{W}_2^T P \\
0 & Q - \alpha I & 0 \\
P\bar{W}_2 & 0 & -\frac{1}{2} P
\end{matrix} \; \prec 0, \tag{5.80}
$$

along with (5.96).

proof The problem of $\mathcal{H}_\infty$ observer design [4, 7] is to determine the gain matrix K_3 such that

$$
\lim_{k \to \infty} e_{f,k} = 0 \quad \text{for } w_k = 0, \tag{5.81}
$$

$$
\|\varepsilon_f\|_{l_2} \le \xi \|w\|_{l_2} \quad \text{for } w_k \neq 0, \; e_{f,0} = 0. \tag{5.82}
$$

In order to settle the above problem it is sufficient to find a Lyapunov function V_k such that

$$
\Delta V_k + \varepsilon_{f,k}^T \varepsilon_{f,k} - \mu^2 w_k^T w_k - \mu^2 w_{k+1}^T w_{k+1} < 0, \; k = 0, \ldots \infty, \tag{5.83}
$$

where $\Delta V_k = V_{k+1} - V_k$, $\mu > 0$. Indeed, if $w_k = 0$, $(k = 0, \ldots, \infty)$, then (5.83) boils down to

$$
\Delta V_k + \varepsilon_{f,k}^T \varepsilon_{f,k} < 0, \; k = 0, \ldots \infty, \tag{5.84}
$$

and hence $\Delta V_k < 0$, which leads to (5.81). If $w_k \neq 0$ $(k = 0, \ldots, \infty)$, then (5.83) yields

$$
J = \sum_{k=0}^{\infty} \left(\Delta V_k + \varepsilon_{f,k}^T \varepsilon_{f,k} - \mu^2 w_k^T w_k - \mu^2 w_{k+1}^T w_{k+1} \right) < 0, \tag{5.85}
$$

which can be written as

$$
J = -V_0 + \sum_{k=0}^{\infty} \varepsilon_{f,k}^T \varepsilon_{f,k} - \mu^2 \sum_{k=0}^{\infty} w_k^T w_k - \mu^2 \sum_{k=0}^{\infty} w_{k+1}^T w_{k+1} < 0. \tag{5.86}
$$

Bearing in mind that

$$
\mu^2 \sum_{k=0}^{\infty} w_{k+1}^T w_{k+1} = \mu^2 \sum_{k=0}^{\infty} w_k^T w_k - \mu^2 w_0^T w_0, \tag{5.87}
$$

the inequality (5.86) can be written as

$$J = -V_0 + \sum_{k=0}^{\infty} \varepsilon_{f,k}^T \varepsilon_{f,k} - 2\mu^2 \sum_{k=0}^{\infty} \boldsymbol{w}_k^T \boldsymbol{w}_k + \mu^2 \boldsymbol{w}_0^T \boldsymbol{w}_0 < 0. \tag{5.88}$$

Knowing that $V_0 = 0$ for $\boldsymbol{e}_{f,0} = 0$, (5.88) leads to (5.82) with $\xi = \sqrt{2}\mu$.

Since the general framework for designing the robust observer is given, the following form of the Lyapunov function is proposed [5]:

$$V_k = \boldsymbol{e}_{f,k}^T \boldsymbol{P} \boldsymbol{e}_{f,k} + \boldsymbol{s}_k^T \boldsymbol{Q} \boldsymbol{s}_k, \tag{5.89}$$

where $\boldsymbol{P} \succ \boldsymbol{0}$ and $\boldsymbol{Q} \succ \boldsymbol{0}$.

Consequently,

$$\begin{aligned}
\Delta V_k &+ \varepsilon_{f,k}^T \varepsilon_{f,k} - \mu^2 \boldsymbol{w}_k^T \boldsymbol{w}_k - \mu^2 \boldsymbol{w}_{k+1}^T \boldsymbol{w}_{k+1} \\
&= \boldsymbol{e}_{f,k}^T \left(\boldsymbol{A}_1^T \boldsymbol{P} \boldsymbol{A}_1 + \boldsymbol{A}^T \boldsymbol{H}_1 \boldsymbol{A} - \boldsymbol{P} \right) \boldsymbol{e}_{f,k} \\
&+ \boldsymbol{e}_{f,k}^T \left(\boldsymbol{A}_1^T \boldsymbol{P} \boldsymbol{G} + \boldsymbol{A}^T \boldsymbol{H}_1 \right) \boldsymbol{s}_k \\
&+ \boldsymbol{e}_{f,k}^T \left(\boldsymbol{A}_1^T \boldsymbol{P} \bar{\boldsymbol{W}}_1 + \boldsymbol{A}^T \boldsymbol{H}_1 \boldsymbol{W}_1 \right) \boldsymbol{w}_k \\
&+ \boldsymbol{e}_{f,k}^T \left(\boldsymbol{A}_1^T \boldsymbol{P} \bar{\boldsymbol{W}}_2 + \boldsymbol{A}^T \boldsymbol{H}_2 \right) \boldsymbol{w}_{k+1} \\
&+ \boldsymbol{s}_k^T \left(\boldsymbol{G}^T \boldsymbol{P} \boldsymbol{A}_1 + \boldsymbol{H}_1 \boldsymbol{A} \right) \boldsymbol{e}_{f,k} \\
&+ \boldsymbol{s}_k^T \left(\boldsymbol{G}^T \boldsymbol{P} \boldsymbol{G} + \boldsymbol{H}_1 - \boldsymbol{Q} \right) \boldsymbol{s}_k \\
&+ \boldsymbol{s}_k^T \left(\boldsymbol{G}^T \boldsymbol{P} \bar{\boldsymbol{W}}_1 + \boldsymbol{H} \boldsymbol{W}_1 \right) \boldsymbol{w}_k \\
&+ \boldsymbol{s}_k^T \left(\boldsymbol{G}^T \boldsymbol{P} \bar{\boldsymbol{W}}_2 + \boldsymbol{H}_2 \right) \boldsymbol{w}_{k+1} \\
&+ \boldsymbol{w}_k^T \left(\bar{\boldsymbol{W}}_1^T \boldsymbol{P} \boldsymbol{A}_1 + \boldsymbol{W}_1^T \boldsymbol{H}_1 \boldsymbol{A} \right) \boldsymbol{e}_{f,k} \\
&+ \boldsymbol{w}_k^T \left(\bar{\boldsymbol{W}}_1^T \boldsymbol{P} \boldsymbol{A}_1 + \boldsymbol{W}_1^T \boldsymbol{H} \right) \boldsymbol{s}_k \\
&+ \boldsymbol{w}_k^T \left(\bar{\boldsymbol{W}}_1 \boldsymbol{P} \bar{\boldsymbol{W}}_1 + \boldsymbol{W}_1^T \boldsymbol{H}_1 \boldsymbol{W}_1 - \mu^2 \boldsymbol{I} \right) \boldsymbol{w}_k \\
&+ \boldsymbol{w}_k^T \left(\bar{\boldsymbol{W}}_1^T \boldsymbol{P} \boldsymbol{W}_2 + \boldsymbol{W}_1^T \boldsymbol{H}_2 \right) \boldsymbol{w}_{k+1} \\
&+ \boldsymbol{w}_{k+1}^T \left(\bar{\boldsymbol{W}}_2^T \boldsymbol{P} \boldsymbol{A}_1 + \boldsymbol{H}_2^T \boldsymbol{A} \right) \boldsymbol{e}_{f,k} \\
&+ \boldsymbol{w}_{k+1}^T \left(\bar{\boldsymbol{W}}_2^T \boldsymbol{P} \boldsymbol{G} + \boldsymbol{H}_2^T \right) \boldsymbol{s}_k \\
&+ \boldsymbol{w}_{k+1}^T \left(\bar{\boldsymbol{W}}_2^T \boldsymbol{P} \boldsymbol{W}_1 + \boldsymbol{H}_2^T \boldsymbol{W}_1 \right) \boldsymbol{w}_k \\
&+ \boldsymbol{w}_{k+1}^T \left(\bar{\boldsymbol{W}}_2^T \boldsymbol{P} \bar{\boldsymbol{W}}_2 + \boldsymbol{W}_2^T \boldsymbol{H}^T \boldsymbol{H} \boldsymbol{W}_2 - \mu^2 \boldsymbol{I} \right) \boldsymbol{w}_{k+1} \\
&+ \boldsymbol{s}_{k+1}^T \boldsymbol{Q} \boldsymbol{s}_{k+1} < 0, \tag{5.90}
\end{aligned}$$

with $H_1 = C^T H^T H C$ and $H_2 = C^T H^T H W_2$.

By defining

$$v_k = \left[e_{f,k}^T,\ s_k^T,\ w_k^T,\ w_{k+1}^T,\ s_{k+1}^T \right]^T, \tag{5.91}$$

the inequality (5.90) becomes

$$\Delta V_k + \varepsilon_{f,k}^T \varepsilon_{f,k} - \mu^2 w_k^T w_k - \mu^2 w_{k+1}^T w_{k+1} = v_k^T M_V v_k < 0, \tag{5.92}$$

where M_V is given by (5.93):

$$M_V = \begin{bmatrix} A_1^T P A_1 - P + A^T H_1 A & A_1^T P G + A_1^T H_1 & A_1^T P \bar{W}_1 \\ +G^T P A_1 + H_1 A & G^T P G + H_1 - Q & G^T P \bar{W}_1 \\ +\bar{W}_1^T P A_1 + W_1^T H_1 A & \bar{W}_1^T P G + W_1^T H_1 & \bar{W}_1^T P \bar{W}_1 \\ +\bar{W}_2^T P A_1 + H_2^T A & \bar{W}_2^T P G + H_2^T & \bar{W}_2^T P \bar{W}_1 \\ \\ +0 & 0 & 0 \end{bmatrix}$$

$$\begin{bmatrix} +A_1^T H W_1 & A_1^T P \bar{W}_2 + A^T H_2 & 0 \\ +H_1 W_1 & G^T P \bar{W}_2 + H_2 & 0 \\ +W_1^T H_1 W_1 - \mu^2 I & \bar{W}_1 P \bar{W}_2 + W_1^T H_2 & 0 \\ +H_2^T W_1 & W_2^T H^T H W_2^T + \bar{W}_2^T P \bar{W}_2 - \mu^2 I & 0 \\ 0 & Q \end{bmatrix}. \tag{5.93}$$

Additionally, from (5.5),

$$\beta e_{f,k} M^T M e_{f,k} - \beta s_k^T s_k \geq 0, \quad \beta > 0, \tag{5.94}$$

which is equivalent to

$$v_k^T \begin{bmatrix} \beta M^T M & 0 & 0 & 0 & 0 \\ 0 & -\beta I & 0 & 0 & 0 \\ 0 & 0 & 0 & 0 & 0 \\ 0 & 0 & 0 & 0 & 0 \\ 0 & 0 & 0 & 0 & 0 \end{bmatrix} v_k \geq 0. \tag{5.95}$$

Similarly, from (5.5) and by assuming that

$$P \succ \alpha M^T M, \quad \alpha > 0, \tag{5.96}$$

it can be shown that

$$\alpha e_{f,k+1}^T M^T M e_{f,k+1} - \alpha s_{k+1}^T s_{k+1}$$
$$< e_{f,k+1}^T P e_{f,k+1} - \alpha s_{k+1}^T s_{k+1} \geq 0, \tag{5.97}$$

which is equivalent to

$$v_k^T M_Y v_k \geq 0, \tag{5.98}$$

with

$$
M_Y = \begin{bmatrix}
A_1^T P A_1 & A_1^T P G & A_1^T P \bar{W}_1 \\
G^T P A_1 & G^T P G & G^T P \bar{W}_1 \\
W^T G^T P A_1 & W^T G^T P G & W^T G^T P G W \\
\bar{W}_1^T P A_1 & \bar{W}_1^T P G & \bar{W}_1^T P \bar{W}_1 \\
\bar{W}_2^T P A_1 & \bar{W}_2^T P G & \bar{W}_2^T P \bar{W}_1 \\
0 & 0 & 0
\end{bmatrix}
$$

$$
\begin{matrix}
A_1^T P \bar{W}_2 & 0 \\
G^T P \bar{W}_2 & 0 \\
0 & 0 \\
\bar{W}_1 P \bar{W}_2 & 0 \\
W_2^T H^T H W_2^T & 0 \\
0 & -\alpha I
\end{matrix}\Bigg].
$$

(5.99)

Finally, using (5.92) along with (5.95) and (5.98), the convergence condition of the observer becomes (5.100):

$$
\begin{bmatrix}
2A_1^T P A_1 - P + A^T H_1 A + \beta M^T M & 2A_1^T P G + A_1^T H_1 \\
2G^T P A_1 + 2H_1 A & G^T P G + H_1 - Q - \beta I \\
2\bar{W}_1^T P A_1 + W_1^T H_1 A & 2\bar{W}_1^T P G + W_1^T H_1 \\
2\bar{W}_2^T P A_1 + H_2^T A & 2\bar{W}_2^T P G + H_2^T \\
\\
0 & 0
\end{bmatrix}
$$

(5.100)

$$
\begin{matrix}
2A_1^T P \bar{W}_1 + A_1^T H W_1 & 2A_1^T P \bar{W}_2 + A^T H_2 & 0 \\
2G^T P \bar{W}_1 + H_1 W_1 & 2G^T P \bar{W}_2 + H_2 & 0 \\
2\bar{W}_1^T P \bar{W}_1 + W_1^T H_1 W_1 - \mu^2 I & 2\bar{W}_1 P \bar{W}_2 + W_1^T H_2 & 0 \\
2\bar{W}_2^T P \bar{W}_1 + H_2^T W_1 & 2W_2^T H^T H W_2^T + \bar{W}_2^T P \bar{W}_2 - \mu^2 I & 0 \\
0 & 0 & Q - \alpha I
\end{matrix}\Bigg] \prec 0.
$$

Moreover, by applying the Schur complements, (5.100) can be transformed into (5.101):

$$
\begin{bmatrix}
-P + A^T H_1 A + \beta M^T M & A^T H_1 & A^T H_1 W_1 \\
H_1 A & H_1 - Q - \beta I & H_1 W_1 \\
W_1^T H_1 A & W_1^T H_1 & W_1^T H_1 W_1 - \mu^2 I \\
H_2^T A & H_2^T & H_2^T W_1 \\
\\
0 & 0 & 0 \\
\\
A_1 & G & \bar{W}_1
\end{bmatrix}
$$

(5.101)

$$
\begin{matrix}
A^T H_2 & 0 & A_1^T \\
H_2 & 0 & G^T \\
W_1^T H_2 & 0 & \bar{W}_1^T \\
W_2^T H^T H W_2 - \mu^2 I & 0 & \bar{W}_2^T \\
0 & Q - \alpha I & 0 \\
\bar{W}_2 & 0 & -\frac{1}{2} P^{-1}
\end{matrix}\Bigg] \prec 0.
$$

Multiplying (5.101) from both sites by

$$
\operatorname{diag}(I, I, I, I, P)
$$

(5.102)

and then substituting

$$PA_1 = P\bar{A} - PK_3C = P\bar{A} - NC,$$
$$P\bar{W}_1 = PGW_1 - PK_3W_2 = PGW_1 - NW_2 \tag{5.103}$$

yield (5.80), which completes the proof.

Note that (5.80) is a usual LMI, which can be easily solved, e.g., with MATLAB. Thus, the final design procedure is as follows. Given a prescribed disturbance attenuation level μ, obtain $\alpha > 0$, $\beta > 0$, $P \succ \mathbf{0}$, $Q \succ \mathbf{0}$, N by solving (5.96) and (5.80). Finally, the gain matrix of (5.75) is

$$K_3 = P^{-1}N. \tag{5.104}$$

It can be also observed that the observer design problem can be treated as an minimisation task, i.e.,

$$\mu^* = \min_{\mu>0,\alpha>0,\beta>0,P\succ\mathbf{0},Q\succ\mathbf{0},N} \mu \tag{5.105}$$

under (5.96) and (5.80).

5.2.3 Guaranteed Decay Rate

The objective of the preceding sections was to show that it is possible to design a fault identification scheme in such a way that a prescribed disturbance attenuation level is achieved with respect to the fault estimation error while guaranteeing the convergence of the observer. In this section it will be pointed out how to attain the additional objective, i.e., a guaranteed decay rate of the state estimation error.

Following [8], for $\mathbf{w}_k = \mathbf{0}$ $(k = 0, \ldots, \infty)$ and $\theta > 0$, the decay rate of the state estimation error is defined as

$$\lim_{k\to\infty} e^{\theta k} \|\mathbf{e}_{f,k}\| = 0. \tag{5.106}$$

Suppose that

$$\Delta V_k = V_{k+1} - V_k \leq -(1 - e^{-2\theta})V_k, \tag{5.107}$$

which is equivalent to

$$V_k \leq e^{-2\theta k} V_0. \tag{5.108}$$

Using (5.6), it can be shown that there exists $M_{x,k}$ such that

$$\mathbf{s}_k = M_{x,k}\mathbf{e}_{f,k}. \tag{5.109}$$

With (5.109), the Lyapunov function (5.83) can be expressed as

$$V_k = e_{f,k}^T P e_{f,k} + e_{f,k}^T M_{x,k}^T Q M_{x,k} e_{f,k} = e_{f,k}^T \left[P + M_{x,k}^T Q M_{x,k} \right] e_{f,k}$$
$$= e_{f,k}^T P_k e_{f,k}. \tag{5.110}$$

Since, for $e_{f,k} \neq 0$, $e_{f,k}^T P e_{f,k} > 0$ and $e_{f,k}^T M_{x,k}^T Q M_{x,k} e_{f,k} \geq 0$, it is evident that $P_k \succ 0$. Bearing in mind that

$$\lambda_{\min}(P_k) e_{f,k}^T e_{f,k} \leq e_{f,k}^T P_k e_{f,k} \leq \lambda_{\max}(P_k) e_{f,k}^T e_{f,k}, \tag{5.111}$$

it can be shown that

$$\| e_{f,k} \| \leq \sqrt{\frac{\lambda_{\max}(P_0)}{\lambda_{\min}(P_k)}} e^{-\theta k} \| e_{f,0} \|, \tag{5.112}$$

which is equivalent to (5.106).

Thus, using (5.107), to attain a guaranteed decay rate θ it is needed that

$$\Delta V \leq -(1 - e^{-2\theta}) V_k = -(1 - e^{-2\theta}) \left(e_{f,k}^T P e_{f,k} + s_k^T Q s_k \right), \tag{5.113}$$

which is equivalent to replacing $-P$ and $-Q$ in (5.80) by $-\tau P$ and $-\tau Q$, respectively ($\tau = e^{-2\theta}$, $\tau > 0$), which gives

$$\begin{bmatrix} -\tau P + A^T H_1 A + \beta M^T M & A^T H_1 & A^T H_1 W_1 \\ H_1 A & H_1 - \tau Q - \beta I & H_1 W_1 \\ W_1^T H_1 A & W_1^T H_1 & W_1^T H_1 W_1 - \mu^2 I \\ H_2^T A & H_2^T & H_2^T W_1 \\ 0 & 0 & 0 \\ P\bar{A} - NC & PG & PGW_1 - NW_2 \\ \end{bmatrix}$$

$$\begin{matrix} A^T H_2 & 0 & \bar{A}^T P - C^T N^T \\ H_2 & 0 & G^T P \\ W_1^T H_2 & 0 & W_1^T G^T P - W_2^T N^T \\ W_2^T H^T H W_2 - \mu^2 I & 0 & \bar{W}_2^T P \\ 0 & Q - \alpha I & 0 \\ P\bar{W}_2 & 0 & -\frac{1}{2} P \end{matrix} \Bigg] \prec 0. \tag{5.114}$$

Note that $\theta = -\frac{1}{2} \ln(\tau)$.

Thus, given a prescribed disturbance attenuation level μ, the following optimisation problem can be formulated:

$$\tau^* = \min_{\tau > 0, \alpha > 0, \beta > 0, P \succ 0, Q \succ 0, N} \tau \tag{5.115}$$

under (5.96) and (5.114).

An alternative solution is to minimise μ and τ simultaneously, i.e., given a scalar $0 \le \lambda \le 1$, the optimisation problem is (cf. [8])

$$\left(\tau^*, \mu^*\right) = \min_{\tau > 0, \mu > 0, \alpha > 0, \beta > 0, \boldsymbol{P} > \boldsymbol{0}, \boldsymbol{Q} > \boldsymbol{0}, \boldsymbol{N}} \lambda\tau + (1 - \lambda)\mu \tag{5.116}$$

under (5.96) and (5.114).

5.2.4 Integrated FTC Design

Similarly as in Sect. 5.1, let us consider a reference model given by

$$\boldsymbol{x}_{k+1} = \boldsymbol{A}\boldsymbol{x}_k + \boldsymbol{B}\boldsymbol{u}_k + \boldsymbol{g}(\boldsymbol{x}_k), \tag{5.117}$$

$$\boldsymbol{y}_{k+1} = \boldsymbol{C}\boldsymbol{x}_{k+1}, \tag{5.118}$$

and $\boldsymbol{L} = \boldsymbol{B}$ in (5.68) and (5.69).

The objective is to design the control strategy $\boldsymbol{u}_{f,k}$ for (5.68) and (5.69) such that the tracking error

$$\boldsymbol{e}_k = \boldsymbol{x}_k - \boldsymbol{x}_{f,k} \tag{5.119}$$

will be asymptotically convergent with guarantying the prescribed disturbance attenuation level. To achieve this goal, the following control strategy is proposed:

$$\boldsymbol{u}_{f,k} = -\hat{\boldsymbol{f}}_{k-1} + \boldsymbol{K}_1(\boldsymbol{x}_k - \hat{\boldsymbol{x}}_{f,k}) + \boldsymbol{K}_2\gamma_k + \boldsymbol{u}_k. \tag{5.120}$$

Taking into account the problems with one-step fault prediction (cf. Sect. 5.1), the following assumption is imposed:

$$\hat{\boldsymbol{f}}_k = \hat{\boldsymbol{f}}_{k-1} + \bar{\boldsymbol{v}}_k, \quad \bar{\boldsymbol{v}}_k \in l_2. \tag{5.121}$$

Bearing in mind that all faults present in real systems have a finite value, such an assumption is fully justified. Thus, for convergence analysis, the following form of FTC is used

$$\boldsymbol{u}_{f,k} = -\hat{\boldsymbol{f}}_k + \bar{\boldsymbol{v}}_k + \boldsymbol{K}_1(\boldsymbol{x}_k - \hat{\boldsymbol{x}}_{f,k}) + \boldsymbol{K}_2\gamma_k + \boldsymbol{u}_k, \tag{5.122}$$

Using a similar approach as in Sect. 5.1 and setting $\boldsymbol{K}_2 = \boldsymbol{H}\boldsymbol{C}$, the tracking error becomes

$$\boldsymbol{e}_{k+1} = \boldsymbol{A}_1\boldsymbol{e}_k + (\boldsymbol{B}\boldsymbol{H}\boldsymbol{C}\boldsymbol{A} - \boldsymbol{B}\boldsymbol{K}_1)\boldsymbol{e}_{f,k} + \boldsymbol{G}\boldsymbol{\omega}_k + \tilde{\boldsymbol{W}}_1\bar{\boldsymbol{w}}_k + \tilde{\boldsymbol{W}}_2\bar{\boldsymbol{w}}_{k+1}, \tag{5.123}$$

with $\boldsymbol{K}_2 = \boldsymbol{H}\boldsymbol{C}$, $\boldsymbol{A}_1 = \boldsymbol{A} - \boldsymbol{B}\boldsymbol{K}_1$ and $\boldsymbol{H} = (\boldsymbol{C}\boldsymbol{B})^+$, where

$$\tilde{W}_1 = [B, \ [BHC - I] W_1], \quad \bar{w}_k = \begin{bmatrix} \bar{v}_k \\ w_k \end{bmatrix}, \tag{5.124}$$

$$\tilde{W}_2 = [B, \ BHW_2]. \tag{5.125}$$

Using the same arguments as in Sect. 4.2, the convergence analysis can be relaxed to the following form of the tracking error:

$$e_{k+1} = A_1 e_k + G\omega_k + \tilde{W}_1 \bar{w}_k + \tilde{W}_2 \bar{w}_{k+1}. \tag{5.126}$$

The following theorem constitutes the main result of the present section.

Theorem 5.4 *For a prescribed disturbance attenuation level* $\mu > 0$ *for the tracking error* (5.126), *the* $\mathcal{H}_\infty$ *controller design problem* (5.122) *for the system* (5.68) *and* (5.69) *is solvable if there exist* $\alpha > 0, \ \beta > 0, \ P \succ 0, \ Q \succ 0, \ N$ *such that the following LMI is satisfied:*

$$\begin{bmatrix} I - P + \beta M^T M & 0 & 0 & 0 & 0 & A_1 P \\ 0 & -Q - \beta I & 0 & 0 & 0 & GP \\ 0 & 0 & -\mu^2 I & 0 & 0 & \tilde{W}_1 P \\ 0 & 0 & 0 & -\mu^2 I & 0 & \tilde{W}_2 P \\ 0 & 0 & 0 & 0 & Q - \alpha I & 0 \\ PA_1^T & PG^T & P\tilde{W}_1^T & P\tilde{W}_2 & 0 & -\frac{1}{2}P \end{bmatrix} \prec 0, \tag{5.127}$$

along with (5.96), *where*

$$A_1 P = AP - BK_1 P = AP - BN. \tag{5.128}$$

proof By defining the Lyapunov function

$$V_k = e_k^T P e_k + \omega_k^T Q \omega_k \tag{5.129}$$

and

$$\Delta V_k + e_k^T e_k - \mu^2 \bar{w}_k^T \bar{w}_k - \mu^2 \bar{w}_{k+1}^T \bar{w}_{k+1} < 0, \tag{5.130}$$

as well as using the same line of reasoning as in Theorem 5.3, it is possible derive (5.127), which completes the proof.

Thus, the final design procedure is as follows: Given a prescribed disturbance attenuation level μ, obtain $\alpha > 0, \ \beta > 0, \ P \succ 0, \ Q \succ 0, \ N$ by solving (5.96) and (5.127). Finally, the gain matrix of the FTC controller is

$$K_1 = NP^{-1}. \tag{5.131}$$

5.2.5 Illustrative Example: Fault Estimation

Let us consider a non-linear system

$$x_{f,k+1} = Ax_{f,k} + Bu_{f,k} + g(x_{f,k}) + Lf_k + W_1w_k, \tag{5.132}$$
$$y_{f,k+1} = Cx_{f,k+1} + W_2w_{k+1}, \tag{5.133}$$

with

$$A = \begin{bmatrix} 0.9944 & -0.1203 & -0.4302 \\ 0.0017 & 0.9902 & -0.0747 \\ 0 & 0.8187 & 0 \end{bmatrix}, \quad B = \begin{bmatrix} 0.4252 \\ -0.0082 \\ 0.1813 \end{bmatrix},$$

$$C = \begin{bmatrix} 1 & 0 & 0 \\ 0 & 1 & 0 \end{bmatrix}, \quad L = B, \quad W_1 = 0.05I_n, \quad W_2 = \begin{bmatrix} 0.1 & 0 & 0 \\ 0 & 0.1 & 0 \end{bmatrix}$$

and

$$g(x_{f,k}) = \left[\sqrt{2}\sin(x_{1,k}), \frac{0.6\cos(12x_{2,k})}{x_{2,k}^2 + 10}, 0 \right]^T. \tag{5.134}$$

Since the system is given, it is straightforward to calculate (5.7), and then

$$M_x^T M_x = \mathrm{diag}\left(2\sin^2(x_{1,k}), \left(\frac{-7.2\sin(12x_{2,k})}{x_{2,k}^2 + 10} - \frac{1.2\cos(12x_{2,k})x_{2,k}}{(x_{2,k}^2 + 10)^2} \right)^2, 0 \right). \tag{5.135}$$

Finally, the application of the proposed procedure cf. (5.11) leads to

$$M^T M = \begin{bmatrix} 2 & 0 & 0 \\ 0 & 0.517 & 0 \\ 0 & 0 & 0 \end{bmatrix}. \tag{5.136}$$

As a result of solving the problem (5.105), the following couple was obtained:

$$\mu^* = 0.769, \quad K = \begin{bmatrix} -0.1505 & 0.0519 \\ 0.0123 & 0.9629 \\ -0.7057 & 0.9540 \end{bmatrix}. \tag{5.137}$$

Let the initial condition for the system and the observer be

$$x_{f,0} = [3, 2, 1]^T, \quad \hat{x}_{f,0} = 0, \tag{5.138}$$

while the input and the exogenous disturbance are

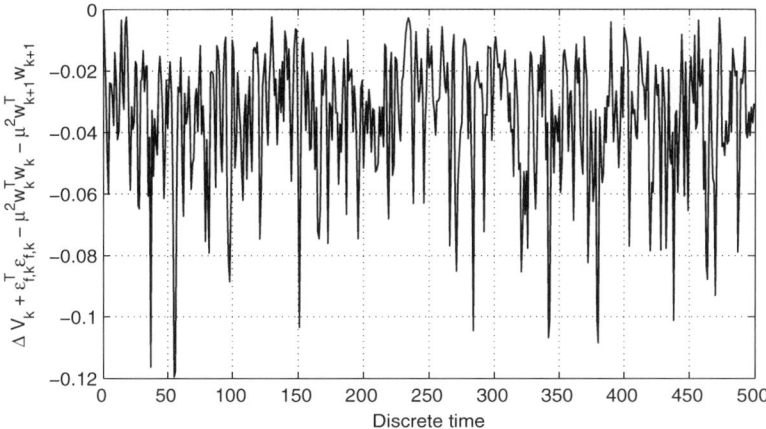

Fig. 5.4 Evolution of $\Delta V_k + \varepsilon_{f,k}^T \varepsilon_{f,k} - \mu^2 \boldsymbol{w}_k^T \boldsymbol{w}_k - \mu^2 \boldsymbol{w}_{k+1}^T \boldsymbol{w}_{k+1}$

$$\boldsymbol{u}_{f,k} = 10, \quad \boldsymbol{w}_k \sim \mathcal{N}(\boldsymbol{0}, 0.1^2 \boldsymbol{I}). \tag{5.139}$$

Moreover, let us consider the following fault scenario:

$$f_k = \begin{cases} 1, & \text{for } 300 \geq k \geq 200, \\ 0, & \text{otherwise.} \end{cases} \tag{5.140}$$

First, let us consider the case when $\hat{\boldsymbol{x}}_{f,0} = \boldsymbol{x}_{f,0}$ ($\boldsymbol{e}_0 = \boldsymbol{0}$). Figure 5.4 clearly indicates that the condition (5.82) is satisfied, which means that an attenuation level $\mu^* = 0.769$ is achieved. Now let us assume that $\boldsymbol{w}_k = \boldsymbol{0}$ and $\hat{\boldsymbol{x}}_{f,0} \neq \boldsymbol{x}_{f,0}$. Figure 5.5 clearly shows that (5.81) is satisfied as well. Finally, Fig. 5.6 shows the fault and its estimate for the nominal case ($\hat{\boldsymbol{x}}_{f,0} \neq \boldsymbol{x}_{f,0}$ and $\boldsymbol{w}_k \neq \boldsymbol{0}$). Let us also consider an incipient fault scenario:

$$f_k = \begin{cases} 0.05(k - 200), & \text{for } 400 \geq k \geq 200, \\ 0, & \text{otherwise.} \end{cases} \tag{5.141}$$

Figure 5.7 presents the attained results, which indicate that also in this case the proposed approach performs well. Now let us assume again that $\boldsymbol{w}_k = \boldsymbol{0}$ and $\hat{\boldsymbol{x}}_{f,0} \neq \boldsymbol{x}_{f,0}$. Moreover, it is assumed that $\mu = 0.99$. As a result of solving the problem (5.115), the following gain matrix was obtained:

$$\boldsymbol{K} = \begin{bmatrix} -0.0744 & 0.0337 \\ 0.0064 & 0.9800 \\ -0.5519 & 0.9255 \end{bmatrix}, \tag{5.142}$$

for which $\tau = 0.9$. Figure (5.8) shows the obtained results. By observing Figs. 5.8 and 5.5, it can be seen that the proposed approach provides an efficient way for increasing

Fig. 5.5 Evolution of $\|e_k\|$ (for $k = 0, \ldots, 20$)

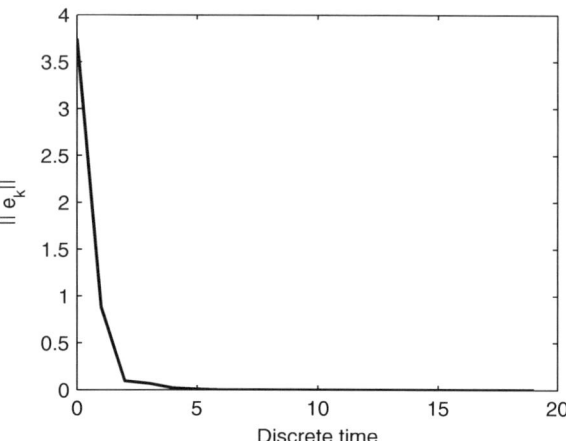

Fig. 5.6 Fault and its estimate

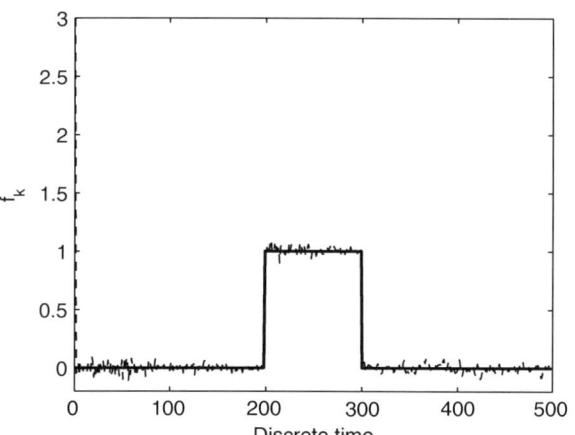

the decay rate of the state estimation error. From these results, it can be observed that the proposed tool can be efficiently applied for solving robust $\mathcal{H}_\infty$-based fault identification of non-linear discrete time systems.

5.2.6 Illustrative Examples: Fault-Tolerant Control

Let us consider the reference model (5.117) and (5.118) with

$$A = \begin{bmatrix} 0.1365 & 0.1991 & 0.2844 \\ 0.0117 & 0.2987 & 0.4692 \\ 0.8939 & 0.6614 & 0.0649 \end{bmatrix}, \quad B = \begin{bmatrix} 1 & 2 \\ 3 & 4 \\ 5 & 6 \end{bmatrix}, \tag{5.143}$$

Fig. 5.7 Incipient fault and its estimate

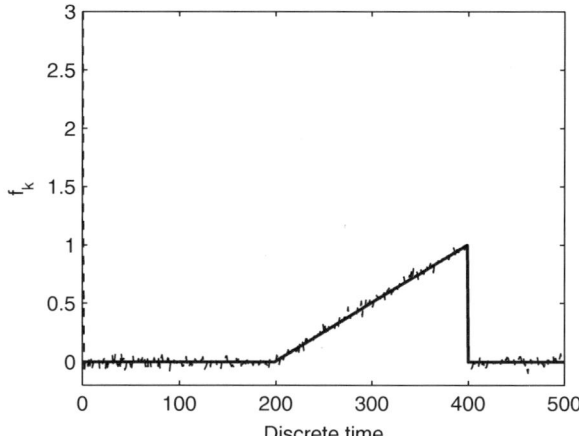

Fig. 5.8 Evolution of $\|e_k\|$ for $\tau = 0.9$ (for $k = 0, \ldots, 20$)

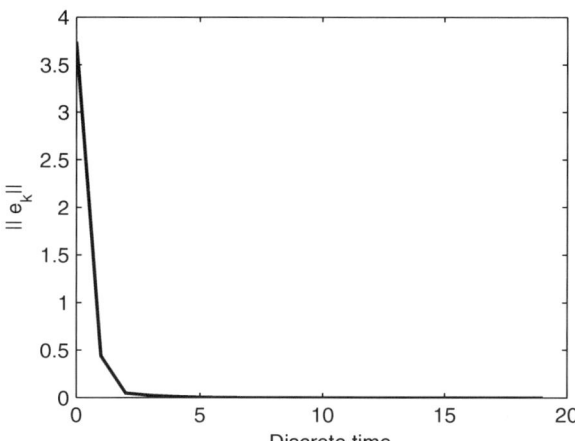

$$C = \begin{bmatrix} 1 & 0 & 0 \\ 0 & 1 & 0 \end{bmatrix}, \quad g(x_k) = [0, 0, -0.7 \sin(x_{3,k})]^T. \tag{5.144}$$

The possibly faulty system (5.68) and (5.69) is described in the same way, and additionally, $L = B$:

$$f_{k,1} = \begin{cases} (0.1 \sin(0.01\pi k) - 0.1) u_{f,k,1}, & 50 < k < 150 \\ 0, & \text{otherwise} \end{cases},$$

$$f_{k,2} = \begin{cases} (0.1 \sin(0.01\pi k) - 0.1) u_{f,k,2}, & 200 < k < 250 \\ 0, & \text{otheriwse} \end{cases},$$

and

$$W_1 = 0.01 I_3, \quad W_2 = \begin{bmatrix} 0.1 & 0 & 0 \\ 0 & 0.1 & 0 \end{bmatrix}, \quad w_k \sim \mathcal{N}(0, 0.1^2 I).$$

Figures 5.9–5.14 present the experimental results for the above-presented fault scenarios. From these results it is clear that in the case of the proposed FTC the state of the possibly faulty system $x_{f,k}$ converges to the reference state x_k irrespective of the faults. It can be also observed that this is not the case for the control strategy without the ability of fault tolerance. Indeed, it can be easily seen that the state x_N diverges from the reference state when faults occur. This fact clearly indicates high performance of the proposed FTC.

5.3 Robust Design for Quasi-LPV Systems

The present section portrays an alternative approach to the one presented in Sect. 5.2. Its main appealing property is that the non-linear system is suitably described in a quasi-LPV form. Subsequently, a robust approach for the estimation of both sensor and actuator faults is presented. It also contains integrated FTC for the actuator faults. The section ends with an example regarding fault estimation of a multi-tank system. Both sensors and actuator faults are considered during the case study.

5.3.1 Actuator Fault Estimation

The main objective of this section is to provide a detailed design procedure for the robust observer, which can be used for actuator fault diagnosis. In other words, the main role of this observer is to provide information about the actuator fault. Indeed, apart from serving as a usual residual generator (see, e.g., [6]), the observer should be designed in such a way that a prescribed disturbance attenuation level is achieved with respect to the actuator fault estimation error while guaranteeing the convergence of the observer.

Let us reconsider a non-linear system

$$x_{f,k+1} = A x_{f,k} + B u_{f,k} + g(x_{f,k}) + L_a f_{a,k} + W_1 w_k, \tag{5.145}$$

$$y_{f,k+1} = C x_{f,k+1} + L_s f_{s,k} + W_2 w_{k+1}, \tag{5.146}$$

with the actuator $f_{a,k}$ and sensor $f_{s,k}$ faults, respectively. Following [2, 6], let us assume that the system is observable and the following rank condition is satisfied:

$$\mathrm{rank}(C L_a) = \mathrm{rank}(L_a) = s. \tag{5.147}$$

Fig. 5.9 State I for fault $f_{k,1}$: x_1, $x_{f,1}$, and $x_{N,1}$—without FTC

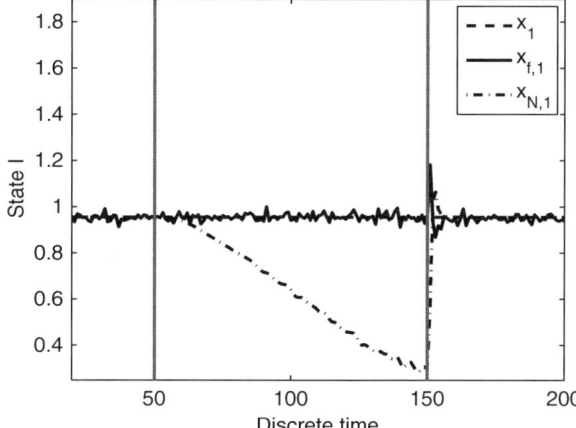

Fig. 5.10 State II for fault $f_{k,1}$: x_2, $x_{f,2}$, and $x_{N,2}$—without FTC

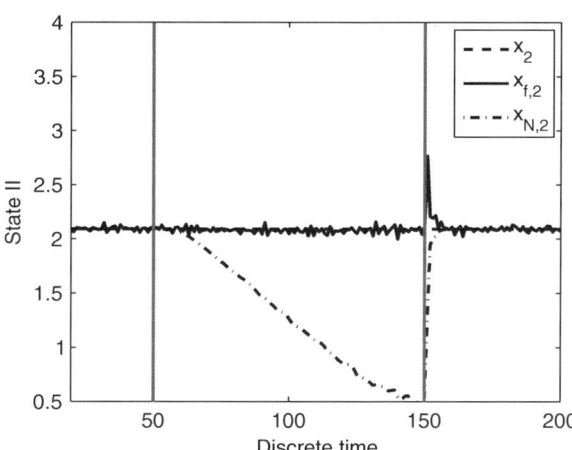

Under the assumption (5.147), it is possible to calculate

$$H = (CL_a)^+ = \left[(CL_a)^T CL_a \right]^{-1} (CL_a)^T. \qquad (5.148)$$

Substituting $f_{s,k} = 0$ into (5.146) as well as multiplying it by H, and then substituting (5.145), it can be shown that

$$\begin{aligned}
f_{a,k} = &H(y_{f,k+1} - CAx_{f,k} - CBu_{f,k} - Cg(x_{f,k}) - CW_1 w_k \\
&- W_2 w_{k+1}).
\end{aligned} \qquad (5.149)$$

Finally, by substituting (5.149) into (5.145) it can be shown that

Fig. 5.11 State III for fault
$f_{k,1}$: x_3, $x_{f,3}$, and $x_{N,3}$—
without FTC

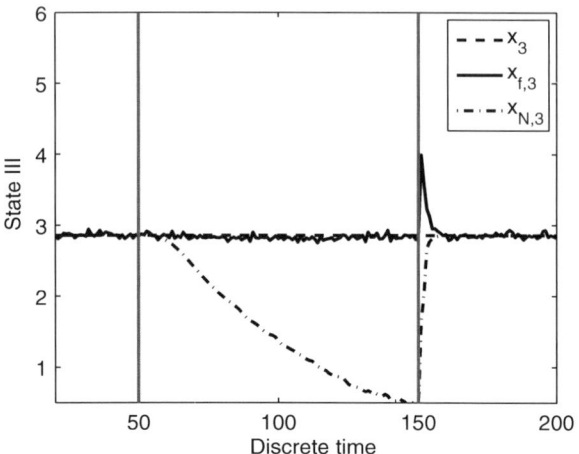

Fig. 5.12 State I for fault
$f_{k,2}$: x_1, $x_{f,1}$, and $x_{N,1}$—
without FTC

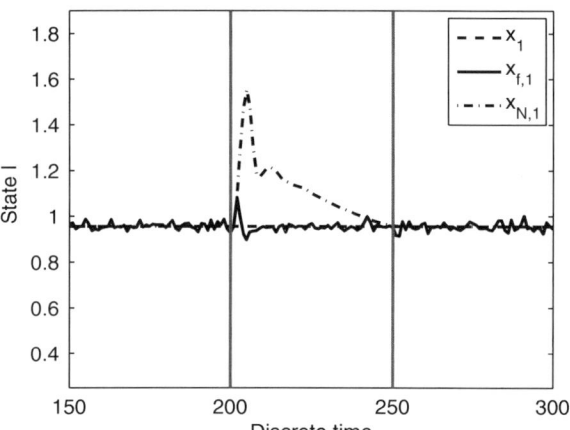

$$x_{f,k+1} = \bar{A}x_{f,k} + \bar{B}u_{f,k} + Gg(x_{f,k}) + \bar{L}y_{f,k+1} + GW_1w_k$$
$$- \bar{L}W_2w_{k+1}, \tag{5.150}$$

where $G = (I_n - L_aHC)$, $\bar{A} = GA$, $\bar{B} = GB$, $\bar{L} = L_aH$. In order to estimate
(5.149), i.e., to obtain $\hat{f}_k$, it is necessary to estimate the state of the system, i.e., to
obtain $\hat{x}_k$. Consequently, the fault estimate is given as follows:

$$\hat{f}_{a,k} = H(y_{f,k+1} - CA\hat{x}_{f,k} - CBu_{f,k} - Cg(\hat{x}_{f,k})). \tag{5.151}$$

The corresponding observer structure is

$$\hat{x}_{f,k+1} = \bar{A}\hat{x}_{f,k} + \bar{B}u_{f,k} + Gg(\hat{x}_{f,k}) + \bar{L}y_{f,k+1} + K_3(y_{f,k} - C\hat{x}_{f,k}), \tag{5.152}$$

Fig. 5.13 State II for fault $f_{k,2}$: x_2, $x_{f,2}$, and $x_{N,2}$—without FTC

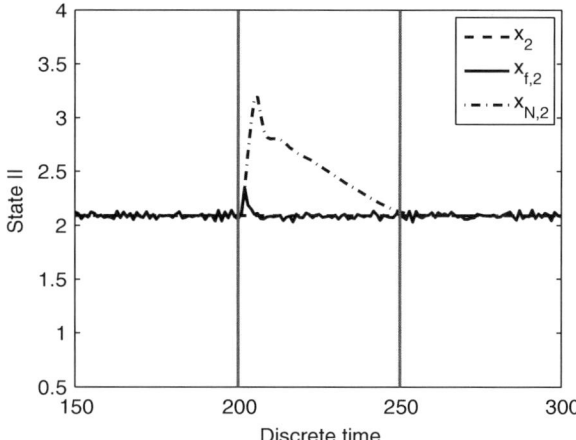

Fig. 5.14 State III for fault $f_{k,2}$: x_3, $x_{f,3}$, and $x_{N,3}$—without FTC

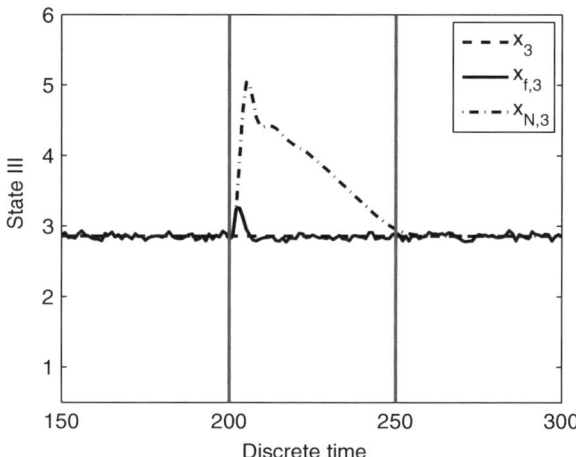

while the state estimation error is given by

$$
\begin{aligned}
e_{f,k+1} &= \left(\bar{A} - K_3 C\right) e_{f,k} + G s_k + (G W_1 - K_3 W_2) w_k - \bar{L} W_2 w_{k+1} \\
&= A_1 e_{f,k} + G s_k + \bar{W}_1 w_k + \bar{W}_2 w_{k+1},
\end{aligned}
\tag{5.153}
$$

where

$$
s_k = g(x_{f,k}) - g(\hat{x}_{f,k}).
\tag{5.154}
$$

Similarly, the fault estimation error $\varepsilon_{f_a,k}$ can be defined as

$$
\varepsilon_{f_a,k} = f_{a,k} - \hat{f}_{a,k} = -HC\left(A e_{f,k} + s_k + W_1 w_k\right) - H W_2 w_{k+1}.
\tag{5.155}
$$

Noth that both $e_{f,k}$ and $\varepsilon_{f_a,k}$ are non-linear with respect to $e_{f,k}$. To settle this problem within the framework of this paper, the following solution is proposed.

Using the DMTV [5], it can be shown that

$$g(a) - g(b) = M_x(a - b), \tag{5.156}$$

with

$$M_x = \begin{bmatrix} \dfrac{\partial g_1}{\partial x}(c_1) \\ \vdots \\ \dfrac{\partial g_n}{\partial x}(c_n) \end{bmatrix}, \tag{5.157}$$

where $c_1, \ldots, c_n \in \mathrm{Co}(a, b)$, $c_i \neq a$, $c_i \neq b$, $i = 1, \ldots, n$. Assuming that

$$\bar{a}_{i,j} \geq \frac{\partial g_i}{\partial x_j} \geq \underline{a}_{i,j}, \quad i = 1, \ldots, n, \quad j = 1, \ldots, n, \tag{5.158}$$

it is clear that

$$\mathbb{M}_x = \left\{ M \in \mathbb{R}^{n \times n} | \bar{a}_{i,j} \geq m_{x,i,j} \geq \underline{a}_{i,j}, \; i, j = 1, \ldots, n, \right\}. \tag{5.159}$$

Thus, using (5.156), the term $A_1 e_{f,k} + G s_k$ in (5.153) can be written as

$$A_1 e_{f,k} + s_k = (\bar{A} + G M_{x,k} - K_3 C) e_{f,k}, \tag{5.160}$$

where $M_{x,k} \in \mathbb{M}_x$.

From (5.160), it can be deduced that the state estimation error can be converted into an equivalent form,

$$e_{f,k+1} = A_2(\alpha) e_{f,k} + \bar{W}_1 w_k + \bar{W}_2 w_{k+1}, \tag{5.161}$$
$$A_2(\alpha) = \tilde{A}(\alpha) - K_3 C,$$

which defines an LPV polytopic system [9] with

$$\tilde{\mathbb{A}} = \left\{ \tilde{A}(\alpha) : \quad \tilde{A}(\alpha) = \sum_{i=1}^{N} \alpha_i \tilde{A}_i, \; \sum_{i=1}^{N} \alpha_i = 1, \; \alpha_i \geq 0 \right\}, \tag{5.162}$$

where $N = 2^{n^2}$. Note that this is a general description, which does not take into account that some elements of $M_{x,k}$ may be constant. In such cases, N is given by $N = 2^{(n-c)^2}$, where c stands for the number of constant elements of $M_{x,k}$.

In a similar fashion, (5.155) can be converted into

$$\varepsilon_{fa,k} = -HC\left(A_3(\alpha)e_{f,k} + W_1 w_k\right) - H W_2 w_{k+1}, \qquad (5.163)$$

with

$$\mathbb{A}_3 = \left\{ A_3(\alpha): \quad A_3(\alpha) = \sum_{i=1}^{N} \alpha_i A_{3,i}, \ \sum_{i=1}^{N} \alpha_i = 1, \ \alpha_i \geq 0 \right\}. \qquad (5.164)$$

The objective of further deliberations is to design the observer (5.152) in such a way that the state estimation error $e_{f,k}$ is asymptotically convergent and the following upper bound is guaranteed:

$$\|\varepsilon_f\|_{l_2} \leq \xi \|\mathbf{w}\|_{l_2}, \qquad (5.165)$$

where $\xi > 0$ is a prescribed disturbance attenuation level. Thus, contrary to the approaches presented in the literature, μ should be achieved with respect to the fault estimation error but not the state estimation error.

Thus, the problem of $\mathcal{H}_\infty$ observer design [4, 7] is to determine the gain matrix K_3 such that

$$\lim_{k \to \infty} e_{f,k} = \mathbf{0} \quad \text{for } \mathbf{w}_k = \mathbf{0}, \qquad (5.166)$$

$$\|\varepsilon_f\|_{l_2} \leq \xi \|\mathbf{w}\|_{l_2} \quad \text{for } \mathbf{w}_k \neq \mathbf{0}, \ e_{f,0} = \mathbf{0}. \qquad (5.167)$$

In order to settle the above problem, it is sufficient to find a Lyapunov function V_k such that

$$\Delta V_k + \varepsilon_{fa,k}^T \varepsilon_{fa,k} - \mu^2 \mathbf{w}_k^T \mathbf{w}_k - \mu^2 \mathbf{w}_{k+1}^T \mathbf{w}_{k+1} < 0, \ k = 0, \ldots \infty, \qquad (5.168)$$

where $\Delta V_k = V_{k+1} - V_k$, $\mu > 0$. Indeed, if $\mathbf{w}_k = \mathbf{0}$, $(k = 0, \ldots, \infty)$ then (5.168) boils down to

$$\Delta V_k + \varepsilon_{fa,k}^T \varepsilon_{fa,k} < 0, \ k = 0, \ldots \infty, \qquad (5.169)$$

and hence $\Delta V_k < 0$, which leads to (5.166). If $\mathbf{w}_k \neq \mathbf{0}$ $(k = 0, \ldots, \infty)$, then (5.168) yields

$$J = \sum_{k=0}^{\infty} \left(\Delta V_k + \varepsilon_{fa,k}^T \varepsilon_{fa,k} - \mu^2 \mathbf{w}_k^T \mathbf{w}_k - \mu^2 \mathbf{w}_{k+1}^T \mathbf{w}_{k+1} \right) < 0, \qquad (5.170)$$

which can be written as

$$J = -V_0 + \sum_{k=0}^{\infty} \varepsilon_{fa,k}^T \varepsilon_{fa,k} - \mu^2 \sum_{k=0}^{\infty} \mathbf{w}_k^T \mathbf{w}_k - \mu^2 \sum_{k=0}^{\infty} \mathbf{w}_{k+1}^T \mathbf{w}_{k+1} < 0. \qquad (5.171)$$

Bearing in mind that

$$\mu^2 \sum_{k=0}^{\infty} \boldsymbol{w}_{k+1}^T \boldsymbol{w}_{k+1} = \mu^2 \sum_{k=0}^{\infty} \boldsymbol{w}_k^T \boldsymbol{w}_k - \mu^2 \boldsymbol{w}_0^T \boldsymbol{w}_0, \tag{5.172}$$

the inequality (5.171) can be written as

$$J = -V_0 + \sum_{k=0}^{\infty} \boldsymbol{\varepsilon}_{fa,k}^T \boldsymbol{\varepsilon}_{fa,k} - 2\mu^2 \sum_{k=0}^{\infty} \boldsymbol{w}_k^T \boldsymbol{w}_k + \mu^2 \boldsymbol{w}_0^T \boldsymbol{w}_0 < 0. \tag{5.173}$$

Knowing that $V_0 = 0$ for $\boldsymbol{e}_{f,0} = 0$, (5.173) leads to (5.167) with $\xi = \sqrt{2}\mu$.

Since the general framework for designing the robust observer is given, then the following form of the Lyapunov function is proposed [5]:

$$V_k = \boldsymbol{e}_{f,k}^T \boldsymbol{P}(\alpha) \boldsymbol{e}_{f,k}, \tag{5.174}$$

where $\boldsymbol{P}(\alpha) \succ \boldsymbol{0}$. Contrary to the design approach presented in the literature (see, e.g. [4] and the references therein), it is not assumed that $\boldsymbol{P}(\alpha) = \boldsymbol{P}$ is constant. Indeed, $\boldsymbol{P}(\alpha)$ can be perceived as a parameter-depended matrix of the form (cf. [9])

$$\boldsymbol{P}(\alpha) = \sum_{i=1}^{N} \alpha_i \boldsymbol{P}_i. \tag{5.175}$$

As a consequence,

$$\begin{aligned}
\Delta V_k + \boldsymbol{\varepsilon}_{fa,k}^T & \boldsymbol{\varepsilon}_{fa,k} - \mu^2 \boldsymbol{w}_k^T \boldsymbol{w}_k - \mu^2 \boldsymbol{w}_{k+1}^T \boldsymbol{w}_{k+1} \\
&= \boldsymbol{e}_{f,k}^T \left(\boldsymbol{A}_2(\alpha)^T \boldsymbol{P}(\alpha) \boldsymbol{A}_2(\alpha) + \boldsymbol{A}_3(\alpha)^T \boldsymbol{H}_1 \boldsymbol{A}_3(\alpha) - \boldsymbol{P}(\alpha) \right) \boldsymbol{e}_{f,k} \\
&\quad + \boldsymbol{e}_{f,k}^T \left(\boldsymbol{A}_2(\alpha)^T \boldsymbol{P}(\alpha) \bar{\boldsymbol{W}}_1 + \boldsymbol{A}_3(\alpha)^T \boldsymbol{H}_1 \boldsymbol{W}_1 \right) \boldsymbol{w}_k \\
&\quad + \boldsymbol{e}_{f,k}^T \left(\boldsymbol{A}_2(\alpha)^T \boldsymbol{P}(\alpha) \bar{\boldsymbol{W}}_2 + \boldsymbol{A}_3(\alpha)^T \boldsymbol{H}_2 \right) \boldsymbol{w}_{k+1} \\
&\quad + \boldsymbol{w}_k^T \left(\bar{\boldsymbol{W}}_1^T \boldsymbol{P}(\alpha) \boldsymbol{A}_2(\alpha) + \boldsymbol{W}_1^T \boldsymbol{H}_1 \boldsymbol{A}_3(\alpha) \right) \boldsymbol{e}_{f,k} \\
&\quad + \boldsymbol{w}_k^T \left(\bar{\boldsymbol{W}}_1^T \boldsymbol{P}(\alpha) \bar{\boldsymbol{W}}_1 + \boldsymbol{W}_1^T \boldsymbol{H}_1 \boldsymbol{W}_1 - \mu^2 \boldsymbol{I} \right) \boldsymbol{w}_k \\
&\quad + \boldsymbol{w}_k^T \left(\bar{\boldsymbol{W}}_1^T \boldsymbol{P}(\alpha) \boldsymbol{W}_2 + \boldsymbol{W}_1^T \boldsymbol{H}_2 \right) \boldsymbol{w}_{k+1} \\
&\quad + r \boldsymbol{w}_{k+1}^T \left(\bar{\boldsymbol{W}}_2^T \boldsymbol{P}(\alpha) \boldsymbol{A}_{2,k} + \boldsymbol{H}_2^T \boldsymbol{A}_3(\alpha) \right) \boldsymbol{e}_{f,k} \\
&\quad + \boldsymbol{w}_{k+1}^T \left(\bar{\boldsymbol{W}}_2^T \boldsymbol{P}(\alpha) \boldsymbol{W}_1 + \boldsymbol{H}_2^T \boldsymbol{W}_1 \right) \boldsymbol{w}_k \\
&\quad + \boldsymbol{w}_{k+1}^T \left(\bar{\boldsymbol{W}}_2^T \boldsymbol{P}(\alpha) \bar{\boldsymbol{W}}_2 + \boldsymbol{W}_2^T \boldsymbol{H}^T \boldsymbol{H} \boldsymbol{W}_2 - \mu^2 \boldsymbol{I} \right) \boldsymbol{w}_{k+1} < \boldsymbol{0}, \tag{5.176}
\end{aligned}$$

with $\boldsymbol{H}_1 = \boldsymbol{C}^T \boldsymbol{H}^T \boldsymbol{H} \boldsymbol{C}$ and $\boldsymbol{H}_2 = \boldsymbol{C}^T \boldsymbol{H}^T \boldsymbol{H} \boldsymbol{W}_2$.

By defining

$$v_k = \left[e_{f,k}^T, \ w_k^T, \ w_{k+1}^T \right]^T,$$ (5.177)

the inequality (5.176) becomes

$$\Delta V_k + \varepsilon_{f_a,k}^T \varepsilon_{f_a,k} - \mu^2 w_k^T w_k - \mu^2 w_{k+1}^T w_{k+1} = v_k^T M_V v_k < 0,$$ (5.178)

where M_V is given by (5.179):

$$M_V = \begin{bmatrix} A_2(\alpha)^T P(\alpha) A_2(\alpha) + A_3(\alpha)^T H_1 A_3(\alpha) - P(\alpha) & A_2(\alpha)^T P(\alpha) \bar{W}_1 + A_3(\alpha)^T H_1 W_1 \\ \bar{W}_1^T P(\alpha) A_2(\alpha) + W_1^T H_1 A_3(\alpha) & \bar{W}_1^T P(\alpha) \bar{W}_1 + W_1^T H_1 W_1 - \mu^2 I \\ \bar{W}_2^T P(\alpha) A_2(\alpha) + H_2^T A_3(\alpha) & \bar{W}_2^T P(\alpha) W_1 + H_2^T W_1 \end{bmatrix}$$

$$\left. \begin{array}{c} A_2(\alpha)^T P(\alpha) \bar{W}_2 + A_3(\alpha)^T H_2 \\ \bar{W}_1^T P(\alpha) W_2 + W_1^T H_2 \\ \bar{W}_2^T P(\alpha) \bar{W}_2 + W_2^T H^T H W_2 - \mu^2 I \end{array} \right].$$ (5.179)

The following theorem constitutes the main result of this section.

Theorem 5.5 *For a prescribed disturbance attenuation level* $\mu > 0$ *for the fault estimation error* (5.155), *the* $\mathcal{H}_\infty$ *observer design problem for the system* (5.145) *and* (5.146) *and the observer* (5.152) *is solvable if there exist matrices* $P_i \succ 0$ *($i = 1, \ldots, N$), U and N such that the following LMIsare satisfied:*

$$\begin{bmatrix} A_{3,i}^T H_1 A_{3,i} - P_i & A_{3,i}^T H_1 W_1 & A_{3,i}^T H_3 & A_{2,i} U^T \\ W_1^T H_1 A_{3,i} & W_1^T H_1 W_1 - \mu^2 I & W_1^T H_2 & \bar{W}_1^T U^T \\ H_2^T A_{3,i} & H_2^T W_1 & W_2^T H^T H W_2 - \mu^2 I & \bar{W}_2^T U^T \\ U A_{2,i} & U \bar{W}_1 & U \bar{W}_2 & P_i - U - U^T \end{bmatrix} \prec 0,$$

$$i = 1, \ldots, N,$$ (5.180)

where (cf. (5.161) *and* (5.153))

$$U A_{2,i} = U(\tilde{A}_i - K_3 C) = U \tilde{A}_i - N C,$$ (5.181)
$$U \bar{W}_1 = U(G W_1 - K_3 W_2) = U G W_1 - N W_2.$$ (5.182)

proof The following two lemmas can be perceived as the generalisation of those presented in [9].

Lemma 5.6 *The following statements are equivalent:*

1. There exists $X \succ 0$ such that

$$V^T X V - W \prec 0.$$ (5.183)

2. There exists $X \succ 0$ such that

$$\begin{bmatrix} -W & V^T U^T \\ UV & X - U - U^T \end{bmatrix} \prec 0. \tag{5.184}$$

proof Applying the Schur complement to (1) gives

$$V^T U^T (U^T + U - X)^{-1} UV - W \prec 0. \tag{5.185}$$

Substituting $U = U^T = X$ yields

$$V^T X V - W \prec 0. \tag{5.186}$$

Thus, (1) implies (2).

Multiplying (5.184) by $T = \begin{bmatrix} I & V^T \end{bmatrix}$ on the left and by T^T on the left of (5.184) gives (5.183), which means that (2) implies (1) and hence the proof is completed.

Lemma 5.7 *The following statements are equivalent:*

1. *There exists $X(\alpha) \succ 0$ such that*

$$V(\alpha)^T X(\alpha) V(\alpha) - W(\alpha) \prec 0. \tag{5.187}$$

2. *There exists $X(\alpha) \succ 0$ such that*

$$\begin{bmatrix} -W(\alpha) & V(\alpha)^T U^T \\ UV(\alpha) & X(\alpha) - U - U^T \end{bmatrix} \prec 0. \tag{5.188}$$

proof The proof can be realised by following the same line of reasoning as the one of Lemma 5.6.

It is easy to show that (5.188) is satisfied if there exist matrices $X_i \succ 0$ such that

$$\begin{bmatrix} -W_i & V_i^T U^T \\ UV_i & X_i - U - U^T \end{bmatrix} \prec 0, \quad i = 1, \ldots, N. \tag{5.189}$$

Subsequently, observing that the matrix (5.179) must be negative definite and writing it as

$$\begin{bmatrix} A_2(\alpha)^T \\ \bar{W}_1^T \\ \bar{W}_2^T \end{bmatrix} P(\alpha) \begin{bmatrix} A_2(\alpha) & \bar{W}_1 & \bar{W}_2 \end{bmatrix} \tag{5.190}$$

$$+ \begin{bmatrix} A_3(\alpha)^T H_1 A_3(\alpha) - P(\alpha) & A_3(\alpha)^T H_1 W_1 & A_3(\alpha)^T H_3 \\ W_1^T H_1 A_3(\alpha) & W_1^T H_1 W_1 - \mu^2 I & W_1^T H_2 \\ H_2^T A_3(\alpha) & H_2^T W_1 & W_2^T H^T H W_2 - \mu^2 I \end{bmatrix} \prec 0, \tag{5.191}$$

and then applying Lemma 5.7 and (5.189) lead to (5.180), which completes the proof.

Finally, the design procedure boils down to solving the LMIs (5.180) and then (cf. (5.181) and (5.182) $K_3 = U^{-1}N$.

It can be also observed that the observer design problem can be treated as a minimisation task, i.e.,

$$\mu^* = \min_{\mu > 0, \, P_1 \succ 0, \, U, \, N} \mu \qquad (5.192)$$

under (5.180).

5.3.2 Sensor Fault Estimation

The main objective of this section is to provide a detailed design procedure of the robust observer, which can be used for sensor fault diagnosis. In other words, the main role of this observer is to provide information about the sensor fault. Indeed, apart from serving as a usual residual generator (see, e.g. [6]), the observer should be designed in such a way that a prescribed disturbance attenuation level is achieved with respect to the sensor fault estimation error while guaranteeing the convergence of the observer.

Let us define the matrix X, to be partitioned in such a way that

$$X = \begin{bmatrix} x_1^T \\ \vdots \\ x_{n_x}^T \end{bmatrix}, \qquad (5.193)$$

where x_j stands for the jth row of X. Let us also denote X^j as the matrix X without the jth row and y^j as a vector y without the jth element.

Sensor fault diagnosis will be realised by a set of m observers of the form

$$\hat{x}_{k+1} = A\hat{x}_k + g(\hat{x}_k) + K_{3,j}\left(y^j_{f,k} - C^j\hat{x}_k\right), \quad j = 1, \ldots, m, \qquad (5.194)$$

while the jth output (for $L_{s,k} = I$) is described by

$$y_{f,j,k} = c_j^T x_{f,k} + w_{2,j}^T w_k + f_{j,k}. \qquad (5.195)$$

Thus

$$f_{s,j,k} = y_{f,j,k} - c_j^T x_{f,k} - w_{2,j}^T w_k, \qquad (5.196)$$

and the jth fault estimate is

$$\hat{f}_{s,j,k} = y_{f,j,k} - c_j^T \hat{x}_k. \qquad (5.197)$$

The fault estimation error $\varepsilon_{f_{j,k}}$ of the jth sensor is

$$\varepsilon_{f_{j,k}} = \boldsymbol{f}_{s,j,k} - \hat{\boldsymbol{f}}_{s,j,k} = -\boldsymbol{c}_j^T \boldsymbol{x}_{f,k} + \boldsymbol{c}_j^T \hat{\boldsymbol{x}}_k - \boldsymbol{w}_{2,j}^T \boldsymbol{w}_k = -\boldsymbol{c}_j^T \boldsymbol{e}_{f,k} - \boldsymbol{w}_{2,j}^T \boldsymbol{w}_k, \quad (5.198)$$

while the state estimation error (for $\boldsymbol{f}_{a,k} = \boldsymbol{0}$) is

$$\boldsymbol{e}_{f,k+1} = \boldsymbol{A}\boldsymbol{e}_{f,k} + \boldsymbol{s}_k - \boldsymbol{K}_{3,j}\boldsymbol{C}^j \boldsymbol{e}_{f,k} - \boldsymbol{K}_{3,j}\boldsymbol{W}_2\boldsymbol{w}_k + \boldsymbol{W}_1\boldsymbol{w}_k, \quad (5.199)$$

$$\boldsymbol{e}_{f,k+1} = \left(\boldsymbol{A} - \boldsymbol{K}_3^j \boldsymbol{C}^j\right)\boldsymbol{e}_{f,k} + \boldsymbol{s}_k - \bar{\boldsymbol{W}}\boldsymbol{w}_k. \quad (5.200)$$

Using a similar approach as that is Sect. 5.3.1, the state estimation error (5.200) can be described as

$$\boldsymbol{e}_{f,k+1} = \boldsymbol{A}_4(\alpha)\,\boldsymbol{e}_{f,k} - \bar{\boldsymbol{W}}\boldsymbol{w}_k, \quad (5.201)$$

$$\boldsymbol{A}_4(\alpha) = \hat{\boldsymbol{A}}(\alpha) - \boldsymbol{K}_3^j \boldsymbol{C}^j, \quad (5.202)$$

where

$$\hat{\mathbb{A}} = \left\{ \hat{\boldsymbol{A}}(\alpha): \quad \hat{\boldsymbol{A}}(\alpha) = \sum_{i=1}^{N} \alpha_i \hat{\boldsymbol{A}}_i, \ \sum_{i=1}^{N} \alpha_i = 1, \ \alpha_i \geq 0 \right\}. \quad (5.203)$$

Similarly as in Sect. 5.3.1, the general framework for designing the robust observer is

$$\Delta V_k + \varepsilon_{f_{i,k}}^T \varepsilon_{f_{i,k}} - \mu^2 \boldsymbol{w}_k^T \boldsymbol{w}_k < 0, \ k = 0, \dots \infty, \quad (5.204)$$

with

$$V_k = \boldsymbol{e}_{f,k}^T \boldsymbol{P}(\alpha)\boldsymbol{e}_{f,k}. \quad (5.205)$$

Consequently, it can be shown that:

$$\begin{aligned}
\Delta V_k + \varepsilon_{f_{i,k}}^T \varepsilon_{f_{i,k}} &- \mu^2 \boldsymbol{w}_k^T \boldsymbol{w}_k \\
&= \boldsymbol{e}_{f,k}^T \left(\boldsymbol{A}_4(\alpha)^T \boldsymbol{P}(\alpha)\boldsymbol{A}_4(\alpha)\right) \boldsymbol{e}_{f,k} \\
&+ \boldsymbol{e}_{f,k}^T \left(\boldsymbol{A}_4(\alpha)^T \boldsymbol{P}(\alpha)\bar{\boldsymbol{W}}\right) \boldsymbol{w}_k \\
&+ \boldsymbol{w}_k^T \left(\bar{\boldsymbol{W}}^T \boldsymbol{P}(\alpha)\boldsymbol{A}_4(\alpha)\right) \boldsymbol{e}_{f,k} \\
&+ \boldsymbol{w}_k^T \left(\bar{\boldsymbol{W}}^T \boldsymbol{P}(\alpha)\bar{\boldsymbol{W}}\right) \boldsymbol{w}_k < 0. \quad (5.206)
\end{aligned}$$

By defining

$$\boldsymbol{v}_k = \left[\boldsymbol{e}_{f,k}^T, \ \boldsymbol{w}_k^T\right]^T, \quad (5.207)$$

the inequality (5.206) becomes

$$\Delta V_k + \varepsilon_{f_s,k}^T \varepsilon_{f_s,k} - \mu^2 w_k^T w_k = v_k^T M_V v_k < 0, \tag{5.208}$$

where

$$M_V = \begin{bmatrix} A_4(\alpha)^T P(\alpha) A_4(\alpha) - P(\alpha) + c_j c_j^T & A_4(\alpha)^T P(\alpha) \bar{W} + c_j w_{2,j}^T \\ \bar{W}^T P(\alpha) A_4(\alpha) + w_{2,j} c_j^T & \bar{W}^T P(\alpha) \bar{W} + w_{2,j} w_{2,j}^T - \mu^2 I \end{bmatrix}. \tag{5.209}$$

Theorem 5.6 *For a prescribed disturbance attenuation level $\mu > 0$ for the fault estimation error (5.198), the $\mathcal{H}_\infty$ observer design problem for the system (5.145) and (5.146) and the observer (5.194) is solvable if there exist matrices $P_i > 0$ $(i = 1, \ldots, N)$, U and N such that the following LMIs are satisfied:*

$$\begin{bmatrix} -P_i + c_j c_j^T & c_j w_{2,j}^T & A_{4,i}^T U^T \\ w_{2,j} c_j^T & w_{2,j} w_{2,j}^T - \mu^2 I & \bar{W}^T U^T \\ U A_{4,i} & U \bar{W} & P_i - U - U^T \end{bmatrix} \prec 0, \tag{5.210}$$

where

$$U A_{4,i} = U(\hat{A}_i - K_3 C) = U \hat{A}_i - NC. \tag{5.211}$$

proof Observing that the matrix (5.209) must be negative definite and writing it as

$$\begin{bmatrix} A_4(\alpha)^T \\ \bar{W}^T \end{bmatrix} P(\alpha) \begin{bmatrix} A_4(\alpha)^T & \bar{W}^T \end{bmatrix} + \begin{bmatrix} -P(\alpha) + c_j c_j^T & c_j w_{2,j}^T \\ w_{2,j} c_j^T & w_{2,j} w_{2,j}^T - \mu^2 I \end{bmatrix} \prec 0, \tag{5.212}$$

and then applying Lemma 5.7 and (5.189) lead to (5.210), which completes the proof.

5.3.3 Integrated FTC Design

Using the same arguments as in Sect. 5.2, the convergence analysis can be relaxed to the following form of the tracking error:

$$e_{k+1} = A_1 e_k + G\omega_k + \tilde{W}_1 \bar{w}_k + \tilde{W}_2 \bar{w}_{k+1}. \tag{5.213}$$

Similarly as in Sect. 5.3.1, (5.213) can be expressed as

$$e_{k+1} = A_2(\alpha)e_k + \tilde{W}_1 \bar{w}_k + \tilde{W}_2 \bar{w}_{k+1}, \tag{5.214}$$

$$A_2(\alpha) = \tilde{A}(\alpha) - BK_1.$$

The following theorem constitutes the main result of the present section.

Theorem 5.7 *For a prescribed disturbance attenuation level $\mu > 0$ for the tracking error (5.213), the $\mathcal{H}_\infty$ controller design problem (5.122) for the system (5.68) and (5.69) is solvable if there exist $\boldsymbol{P} \succ \boldsymbol{0}$, $\boldsymbol{U}$, $\boldsymbol{V}$ such that the following LMIs are satisfied:*

$$
\begin{bmatrix}
\boldsymbol{I} - \boldsymbol{P}_i & \boldsymbol{0} & \boldsymbol{0} & \boldsymbol{A}_{2,i}\boldsymbol{U} \\
\boldsymbol{0} & -\mu^2\boldsymbol{I} & \boldsymbol{0} & \tilde{\boldsymbol{W}}_1^T\boldsymbol{U}^T \\
\boldsymbol{0} & \boldsymbol{0} & \mu^2\boldsymbol{I} & \tilde{\boldsymbol{W}}_2^T\boldsymbol{U}^T \\
\boldsymbol{U}^T\boldsymbol{A}_{2,i}^T & \boldsymbol{U}\tilde{\boldsymbol{W}}_1 & \boldsymbol{U}\tilde{\boldsymbol{W}}_2 & \boldsymbol{P}_i - \boldsymbol{U} - \boldsymbol{U}^T
\end{bmatrix} \prec \boldsymbol{0}, \quad i = 1, \ldots, N, \tag{5.215}
$$

with

$$
\boldsymbol{A}_{2,i}\boldsymbol{U} = \left(\tilde{\boldsymbol{A}}_i - \boldsymbol{B}\boldsymbol{K}_1\right)\boldsymbol{U} = \tilde{\boldsymbol{A}}_i\boldsymbol{U} - \boldsymbol{B}\boldsymbol{V}. \tag{5.216}
$$

proof The proof is similar to that of Theorem 5.5.

Thus, the final design procedure is as follows: Given a prescribed disturbance attenuation level μ, obtain $\boldsymbol{P} \succ \boldsymbol{0}, \boldsymbol{U}, \boldsymbol{V}$ by solving (5.215). Finally, the gain matrix of the FTC controller is

$$
\boldsymbol{K}_1 = \boldsymbol{V}\boldsymbol{U}^{-1}. \tag{5.217}
$$

5.3.4 Illustrative Example: Fault Estimation of a Multi-Tank System

The multi-tank system considered (Fig. 5.15) is designed for simulating real industrial multi-tank system in laboratory conditions [10]. The multi-tank system can be efficiently used to practically verify both linear and non-linear control, identification and diagnostics methods. The multi-tank system consists of three separate tanks placed one above the other and equipped with drain valves and level sensors based on hydraulic pressure measurement. Each of them has a different cross-section in order to reflect system nonlinearities. The lower bottom tank is a water reservoir for the system. A variable speed water pump is used to fill the upper tank. The water outflows the tanks due to gravity. The multi-tank system considered has been designed to operate with an external, PC-based digital controller. The control computer communicates with the level sensors, valves and a pump by a dedicated I/O board and the power interface. The I/O board is controlled by real-time software, which operates in a Matlab/Simulink environment. The non-linear discrete-time model of the multi-tank system is given as follows (cf. Fig. 5.16):

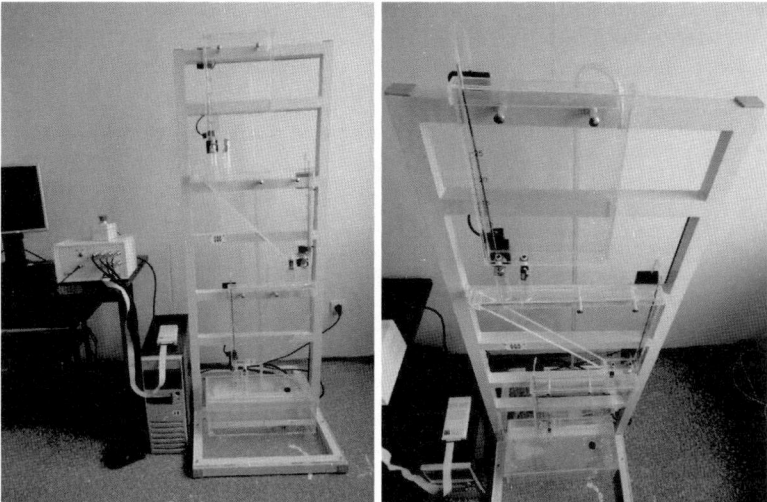

Fig. 5.15 Multi-tank system

$$x_{1,k+1} = x_{1,k} + h\left(\frac{1}{\beta_1(x_{1,k})}u_k - \frac{1}{\beta_1(x_{1,k})}C_1 x_{1,k}^{\alpha_1}\right), \tag{5.218}$$

$$x_{2,k+1} = x_{2,k} + h\left(\frac{1}{\beta_2(x_{2,k})}C_1 x_{1,k}^{\alpha_1} - \frac{1}{\beta_2(x_{2,k})}C_2 x_{2,k}^{\alpha_2}\right), \tag{5.219}$$

$$x_{3,k+1} = x_{3,k} + h\left(\frac{1}{\beta_3(x_{3,k})}C_2 x_{2,k}^{\alpha_2} - \frac{1}{\beta_3(x_{3,k})}C_3 x_{3,k}^{\alpha_3}\right), \tag{5.220}$$

where $x_{i,k}$, $i \in 1, \ldots, 3$, is water level in the ith tank, $\beta_i(x_{i,k})$ stands for the cross section area of the ith tank at the level $x_{i,k}$ and is, respectively, defined as

$\beta_1(x_{1,k}) = aw$: constant cross-sectional area of the top tank;

$\beta_2(x_{2,k}) = cw + \frac{x_{2,k}}{x_{2max}}bw$: variable cross-sectional area of the middle tank;

$\beta_3(x_{3,k}) = w\sqrt{R^2 - (R - x_{3,k})^2}$: variable cross-sectional area of the bottom tank.

The numerical values of the above parameters are as follows: $C_1 = 1.0057 \cdot 10^{-4}$, $C_2 = 1.1963 \cdot 10^{-4}$, $C_3 = 9.8008 \cdot 10^{-5}$, $b = 0.34$, $c = 0.1$, $w = 0.035$, $R = 0.364$, $x_{2max} = 0.35$, $\alpha_1 = 0.29$, $\alpha_2 = 0.2256$, $\alpha_3 = 0.2487$, and $h = 0.01s$.

In order to unify further deliberations, the multi-tank system can be described in a state-space form as:

$$x_{f,k+1} = Ax_{f,k} + Bu_{f,k} + g(x_{f,k}) + L_a f_{a,k} + W_1 w_k, \tag{5.221}$$

$$y_{f,k+1} = Cx_{f,k+1} + L_s f_{s,k} + W_2 w_{k+1}, \tag{5.222}$$

where $x_{f,k} \in \mathbb{X} \subset \mathbb{R}^n$ is the state vector describing the liquid level in the tanks ($n = 3$), $u_{f,k} \in \mathbb{R}^r$ stands for the input ($r = 1$), $y_{f,k} \in \mathbb{R}^m$ denotes the output

Fig. 5.16 Geometrical parameters of the tanks

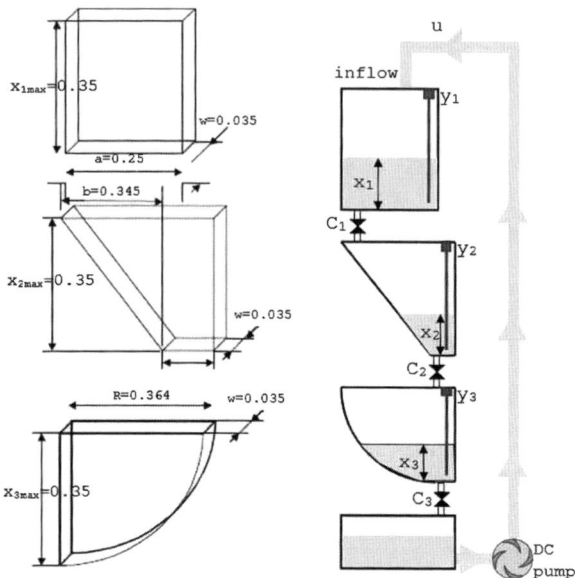

$(m = 3)$, and $\boldsymbol{f}_{a,k} \in \mathbb{R}^s$, $(s = r)$, $\boldsymbol{f}_{s,k} \in \mathbb{R}^m$ stand for the actuator and sensor fault, respectively, while $\boldsymbol{w}_k \in \mathbb{R}^n$ is a an exogenous disturbance vector with $\boldsymbol{W}_1 \in \mathbb{R}^{n \times n}$, $\boldsymbol{W}_2 \in \mathbb{R}^{m \times n}$ being its distribution matrices.

The system matrices and non-linearities are

$$\boldsymbol{A} = \boldsymbol{I}_n, \quad \boldsymbol{B} = \begin{bmatrix} 0.14 \\ 0 \\ 0 \end{bmatrix}, \quad \boldsymbol{C} = \boldsymbol{I}_m,$$

$$\boldsymbol{L}_a = \boldsymbol{B}, \quad \boldsymbol{L}_s = \boldsymbol{I}_m,$$

$$\boldsymbol{g}(\boldsymbol{x}_{f,k}) = \begin{bmatrix} \frac{1}{\beta_1(x_{1,k})} C_1 x_{1,k}^{\alpha_1} \\ \frac{1}{\beta_2(x_{2,k})} C_1 x_{1,k}^{\alpha_1} - \frac{1}{\beta_2(x_{2,k})} C_2 x_{2,k}^{\alpha_2} \\ \frac{1}{\beta_3(x_{3,k})} C_2 x_{2,k}^{\alpha_2} - \frac{1}{\beta_3(x_{3,k})} C_3 x_{3,k}^{\alpha_3} \end{bmatrix}. \tag{5.223}$$

As has already been mentioned, the robustness problem will be tackled within the $\mathcal{H}_\infty$ framework, which yields the following assumption:

$$\boldsymbol{w} \in l_2, \quad l_2 = \left\{ \boldsymbol{w} \in \mathbb{R}^n | \, \|\boldsymbol{w}\|_{l_2} < +\infty \right\}, \quad \|\boldsymbol{w}\|_{l_2} = \left(\sum_{k=0}^{\infty} \|\boldsymbol{w}_k\|^2 \right)^{\frac{1}{2}}. \tag{5.224}$$

Fig. 5.17 Distribution of the disturbances for the top tank level sensor

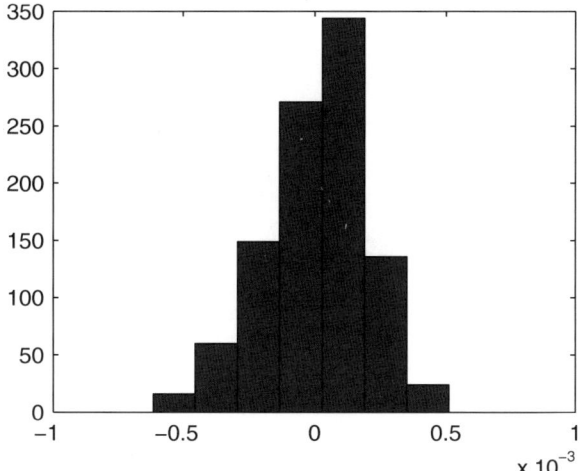

The distribution matrices W_1 and W_2 should express the influence and magnitude of w_k onto the state and output Eqs. (5.145) and (5.146), respectively. To obtain an appropriate proportion between the elements of W_1 and W_2, a series of constant liquid level measurements was performed for the top tank. Subsequently, the mean was removed, which represents the constant liquid level, and then the disturbances were analysed. Figure 5.17 depicts the histogram of the estimated disturbances. The standard deviation of the disturbance is equal to $1.75 \cdot 10^{-4}$ (obtained for 1000 measurements). Almost identical results were obtained for the sensors in the middle and bottom tanks. This is not surprising since all sensors are identical.

The term $W_1 w_k$ (cf. 5.145) will represent the inaccuracy of the pump with respect to a desired control action. After similar experiments like those for the sensors, it was derived that the maximum magnitude of $W_1 w_k$ is approximately 10 times smaller than that of $W_2 w_k$. As a result, the following settings of the distribution matrices were established:

$$W_1 = \mathrm{diag}(0.001, 0, 0), \quad W_2 = 0.01 I_m. \tag{5.225}$$

The analysis was started with the estimation of the actuator fault. To make the experiment more difficult, it was assumed that the measurement of the liquid level of the last tank is not available, which means that

$$C = \begin{bmatrix} 1 & 0 & 0 \\ 0 & 1 & 0 \end{bmatrix}.$$

Let the initial condition for the system and the observer be

$$x_0 = [0.1, 0.2, 0.3]^T, \quad \hat{x}_0 = 1 \cdot 10^{-4}[1, 1, 1]^T, \tag{5.226}$$

Fig. 5.18 Evolution of $\Delta V_k + \varepsilon_{f_a,k}^T \varepsilon_{f_a,k} - \mu^2 \boldsymbol{w}_k^T \boldsymbol{w}_k - \mu^2 \boldsymbol{w}_{k+1}^T \boldsymbol{w}_{k+1}$

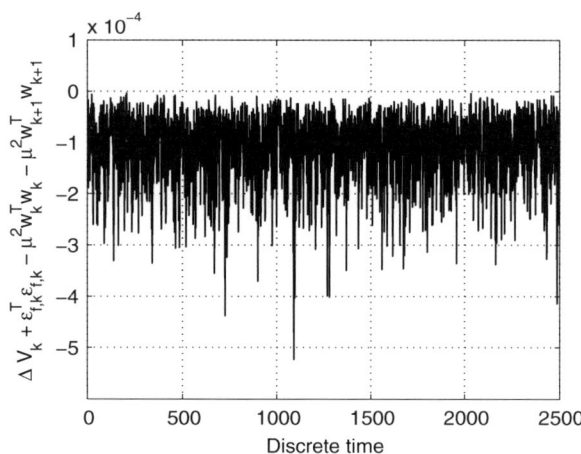

while the input is

$$\boldsymbol{u}_k = 1. \tag{5.227}$$

As a result of solving the problem (5.180), the following couple was obtained:

$$\mu = 0.45; \quad \boldsymbol{K} = \begin{bmatrix} 0.1089 & 0 \\ 0.0004 & 1.7107 \\ 0 & 0.9473 \end{bmatrix}. \tag{5.228}$$

Let us consider the following actuator fault scenario:

$$\boldsymbol{f}_{a,k} = \begin{cases} 0.1, & \text{for } 300 \geq k \geq 200, \\ -0.2\boldsymbol{u}_k & \text{for } 1500 \geq k \geq 500, \\ 0.0005(k - 1800) & \text{for } 2100 \geq k \geq 1800, \\ 0, & \text{otherwise.} \end{cases} \tag{5.229}$$

Figure 5.18 clearly indicates that the condition (5.167) is satisfied, which means that an attenuation level $\mu = 0.45$ is achieved. Finally, Figs. 5.19–5.21 show the faults and their estimates. Moreover, Fig. 5.22 shows the system state $\boldsymbol{x}_{3,k}$ and its estimate $\hat{\boldsymbol{x}}_{3,k}$. The obtained results clearly indicate that the proposed approach works with desired performance.

The objective of the subsequent example is to show the performance of the proposed sensor fault estimation scheme. The example concerns the second sensor's faults. Thus, the following fault scenarios were considered:

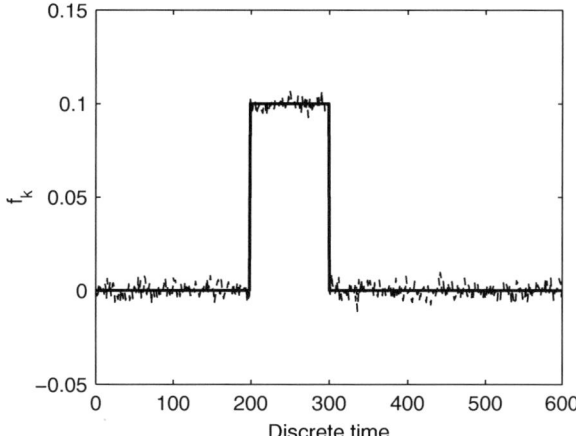

Fig. 5.19 Constant bias actuator fault and its estimate

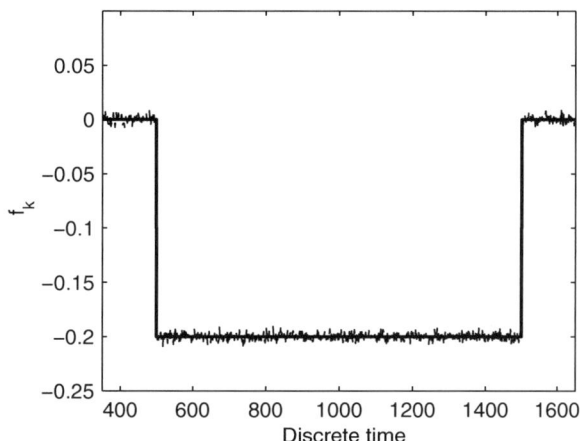

Fig. 5.20 20% decrease fault in actuator and its estimate

1. Incipient fault:

$$f_{s2,k} = \begin{cases} -0.005(k - 1000), & \text{for } 1500 \geq k \geq 1000, \\ 0, & \text{otherwise,} \end{cases}$$

2. 20% decrease in the accuracy:

$$f_{s2,k} = \begin{cases} -0.2y_k, & \text{for } 1500 \geq k \geq 1000, \\ 0, & \text{otherwise.} \end{cases}$$

The results of fault identification are shown in Fig. 5.23. From these it is clear that the proposed approach provides satisfactory outcome. This means that the estimation of

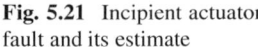

Fig. 5.21 Incipient actuator fault and its estimate

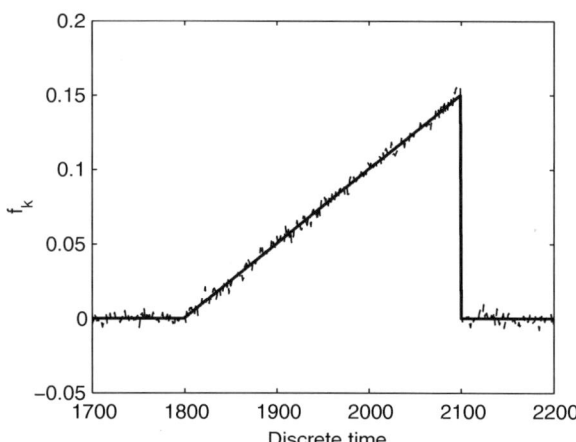

Fig. 5.22 State $x_{3,k}$ and its estimate $\hat{x}_{3,k}$

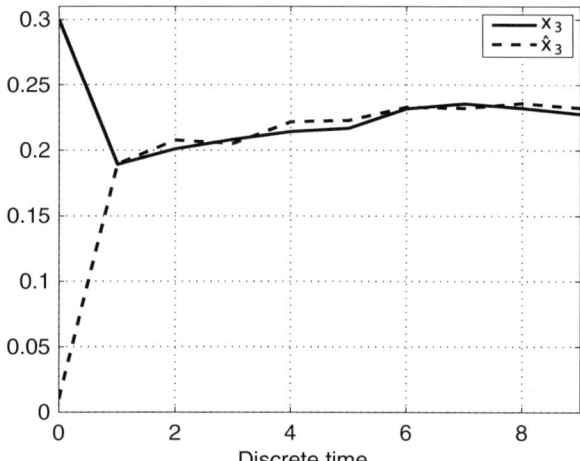

a given sensor measurements can be used in the case when the sensor is faulty. Such an approach will be employed in the subsequent chapter.

5.4 Concluding Remarks

The main objective of this chapter was to propose a novel structure and design procedure of an integrated fault identification and fault-tolerant control scheme for a class of non-linear discrete-time systems. In particular, the chapter was divided into three parts. The first one introduced the concept of integrated fault identification and fault-tolerant control along with robustness with respect to the uncertainties acting

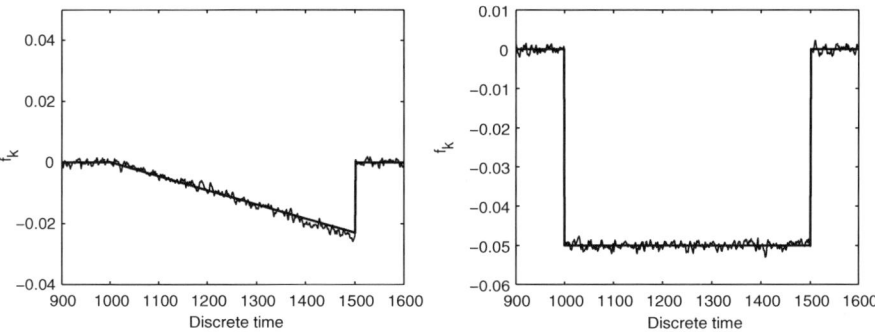

Fig. 5.23 Fault identification for the incipient and the 20 % decrease fault in the second sensor

on the state equation of the system as well as the unappealing effect of one-time step prediction of the fault. The second part enhanced robustness even further. Indeed, additionally, the uncertainties present in the output equation were considered. The last part showed how to implement the proposed FTC framework for LPV systems. It is worth mentioning that fault estimation and control robustness were attained with the $\mathcal{H}_\infty$ approach. In the usual $\mathcal{H}_\infty$ framework, the prescribed disturbance attenuation level is achieved with respect to the state estimation error. The proposed approach is designed in such a way that a prescribed disturbance attenuation level is achieved with respect to the fault estimation error while guaranteeing the convergence of the observer with a possibly large decay rate of the state estimation error. The same property is achieved with respect to the control tracking error. Each part of the chapter had a similar structure, i.e., fault estimation and integrated FTC were carefully analysed. Each was also finalised with the illustrative examples, which showed the performance of the selected solutions.

References

1. D.M. Stipanovic, D.D. Siljak, Robust stability and stabilization of discrete-time non-linear: the lmi approach. Int. J. Control **74**(5), 873–879 (2001)
2. S. Gillijns, B. De Moor, Unbiased minimum-variance input and state estimation for linear discrete-time systems. Automatica **43**, 111–116 (2007)
3. E.G. Nobrega, M.O. Abdalla, K.M. Grigoriadis, Robust fault estimation of unceratain systems using an LMI-based approach. Int. J. Robust Nonlinear Control **18**(7), 1657–1680 (2008)
4. A. Zemouche, M. Boutayeb, G. Iulia Bara, Observer for a class of Lipschitz systems with extension to $\mathcal{H}_\infty$ performance analysis. Syst. Control Lett. **57**(1), 18–27 (2008)
5. A. Zemouche, M. Boutayeb, Observer design for Lipschitz non-linear systems: the discrete time case. IEEE Trans. Circuits Syst. II: Express Briefs **53**(8), 777–781 (2006)
6. M. Witczak, *Modelling and Estimation Strategies for Fault Diagnosis of Non-linear Systems* (Springer-Verlag, Berlin, 2007)
7. H. Li, M. Fu, A linear matrix inequality approach to robust h_∞ filtering. IEEE Trans. Signal Process. **45**(9), 2338–2350 (1997)

8. M. Abbaszadeh, H.J. Marquez, LMI optimization approach to robust $\mathcal{H}_\infty$ observer design and static output feedback stabilization for non-linear uncertain systems. Int. J. Robust Nonlinear Control **19**(3), 313–340 (2008)

9. M.C. de Oliveira, J. Bernussou, J.C. Geromel, A new discrete-time robust stability condition. Syst. Control Lett. **37**(4), 261–265 (1999)

10. INTECO. Multitank System-User's manual. INTECO, www.inteco.com.pl, 2013

Chapter 6
Fuzzy Multiple-Model Approach to Fault-Tolerant Control

6.1 Essentials of Fuzzy Logic

Fuzzy logic is a superset of the conventional (Boolean) logic that was extended to handle the concept of partial truth—truth values between "completely true" and "completely false". It was introduced by Dr. Lotfi Zadeh [1] as a means of the vagueness of a natural language modelling. Initially, it encountered scepticism, and it took a long time until it was finally accepted. Nowadays, fuzzy logic systems are widespread and has found numerous applications, especially in the domain of control engineering, identification, as well as modern computer science.

As there is a strong relationship between Boolean logic and the concept of a subset, there is a similar strong relationship between fuzzy logic and a fuzzy subset theory.

Definition 6.1 A Classical set *F is a set of ordered pairs*

$$F = \{(w, I_F(w)) \mid \forall w \in \mathbb{W}\} \tag{6.1}$$

defined by an indicator function $I_F(w) \in \{0, 1\}$.

The value zero of the indicator function is used to represent non-membership, while the value one is used to represent membership.

For example, if the set of young people F is described as a crisp interval of people younger than, say 20 years, i.e., $F = [0, 20]$. Then the question arises: Why is somebody on his 20th birthday young and right the next day not young? Obviously, this is a structural problem, if the upper bound of the range is moved from 20 to an arbitrary point, the same question can be imposed.

A more natural way to construct the set F would be to relax the strict separation between young and not young. This can be accomplished by allowing not only the crisp decision YES he/she is in the set of young people or NO he/she is not in the set of young people but more flexible phrases like: Well, he/she belongs a little bit more to the set of young people or NO, he/she belongs nearly not to the set of young

M. Witczak, *Fault Diagnosis and Fault-Tolerant Control Strategies for Non-Linear Systems*, 189
Lecture Notes in Electrical Engineering 266, DOI: 10.1007/978-3-319-03014-2_6,
© Springer International Publishing Switzerland 2014

people. Thus, the concept of a *fuzzy set* is introduced, and *young* is described as a *linguistic variable*,[1] which represents humans' cognitive category of "age".

Definition 6.2 A fuzzy set *F* *is a set of ordered pairs*

$$F = \{(w, \mu_F(w)) \mid \forall w \in \mathbb{W}\} \tag{6.2}$$

defined by a membership function $0 \leqslant \mu_F(x) \leqslant 1$.

A membership function provides a measure of the degree of similarity of an element in $\mathbb{W}$ to the fuzzy subset. In practice, the terms 'membership function' and 'fuzzy subset' are used interchangeably.

Now, the idea of fuzzy sets and basic operations on them can be introduced. In the fuzzy logic, union, intersection and complement are defined in terms of membership functions and are motivated by their crisp counterparts. Let fuzzy sets F_1 and F_2 be described by their membership functions $\mu_{F_1}(w)$ and $\mu_{F_2}(w)$. The definition of a *fuzzy intersection* leads to the membership function

$$\mu_{F_1 \cap F_2}(w) = \min[\mu_{F_1}(w), \mu_{F_2}(w)], \quad \forall w \in \mathbb{W}, \tag{6.3}$$

and the definition of a *fuzzy union* leads to the membership function

$$\mu_{F_1 \cup F_2}(w) = \max[\mu_{F_1}(w), \mu_{F_2}(w)] \quad \forall w \in \mathbb{W}. \tag{6.4}$$

Additionally, the membership function for a *fuzzy complement* is

$$\mu_{\bar{F}_1}(w) = 1 - \mu_{F_1}(w) \quad \forall w \in \mathbb{W}. \tag{6.5}$$

The 'max' and 'min' operators are not the only ones that could be chosen to model the fuzzy union and fuzzy intersection. Other operators which have an axiomatic basis can be used, e.g., a *t*-conorm operator for the fuzzy union (also known as an *s*-norm, and denoted by $\mathcal{S}$). Similarly, the bounded sum, drastic sum, and *t*-norm operators can be used for the fuzzy intersection (denoted $\mathcal{T}$) [4].

Since the essential definitions are given, it is possible to proceed to the core structure used within the framework of this chapter, i.e., a Takagi–Sugeno model. It should also be noted that the results presented in this chapter are based on [5].

6.2 Fuzzy Multiple-Model Representation

A non-linear dynamic system can be described in a simple way by a Takagi–Sugeno fuzzy model, being a branch of a general fuzzy framework, which uses series of locally linearised models from the non-linear system, parameter identification of an

---

[1] However, [2] demonstrated that to use a (type 1) fuzzy set to model a word is scientifically incorrect, because word is uncertain whereas a fuzzy set is certain. To do so a type 2 fuzzy set is required, for an example the reader is referred to [3].

a priori given structure or a transformation of a non-linear model using the non-linear sector approach (see, e.g., [6–9]). According to this model, a non-linear dynamic system can be linearised around a number of operating points. Each of these linear models represents the local system behaviour around the operating point. Thus, fuzzy fusion of all linear model states describes global system behaviour.

Let us consider a non-linear model affine in control:

$$\begin{cases} x_{k+1} = g_1(s_k)x_k + g_2(s_k)u_k, \\ y_k = g_3(s_k)x_k, \end{cases} \tag{6.6}$$

with $g_i(\cdot)$, $i = 1, 2, 3$, being non-linear functions and s_k is a vector assumed to be measurable. A methodical way to deal with (6.6) is Takagi–Sugeno modelling [6]. Depending on the 'point of view', two approaches are available leading to a unique framework [10]. From the historical viewpoint, the first approach stems from the fuzzy rule-based control area and its property of being a universal approximator [11]. In this class of fuzzy modelling, the TS fuzzy model acts as an approximation of (6.6), thus allowing description of a non-linear dynamic system by a set of Linear Time Invariant (LTI) models interconnected with non-linear functions. Each of the LTI models is then associated by a rule at the consequent part of a weighting function established based on the premise variable s_k. It has a base of M rules, each having p antecedents, where the ith rule is expressed as (in a state-space representation)

$$R^i : \text{IF } s_k^1 \text{ is } F_1^i \text{ and} \dots \text{ and } s_k^p \text{ is } F_p^i,$$

$$\text{THEN} \begin{cases} x_{k+1}^i = A^i x_k^i + B^i u_k, \\ y_k^i = C^i x_k^i, \end{cases} \tag{6.7}$$

in which $x_k^i \in \mathbb{R}^n$ stands for the state, $y_k^i \in \mathbb{R}^m$ is the output (note that each model had an individual state and output), and $u_k \in \mathbb{R}^r$ denotes the nominal control input, also $i = 1, \dots, M$, F_j^i $(j = 1, \dots, p)$ are fuzzy sets and $s_k = [s_k^1, s_k^2, \dots, s_k^p]$ is a known vector of premise variables [6]. In a general manner, these models are obtained via an identification procedure, according to the universal approximation property [12, 13]. Another approach uses directly the non-linear expression of the model and can be expressed in a rule-based form, although not strictly equivalent to (6.7), where the ith rule is described as

$$R^i : \text{IF } s_k^1 \text{ is } F_1^i \text{ and} \dots \text{ and } s_k^p \text{ is } F_p^i,$$

$$\text{THEN} \begin{cases} x_{k+1} = A^i x_k + B^i u_k, \\ y_k = C^i x_k. \end{cases} \tag{6.8}$$

Given a pair of (s_k, u_k) and a product inference engine, the final output of the normalised TS fuzzy model can be expressed as

$$\begin{cases} x_{k+1} = \sum_{i=1}^{M} h_i(s_k)[A^i x_k + B^i u_k], \\ y_k = \sum_{i=1}^{M} h_i(s_k) C^i x_k, \end{cases} \tag{6.9}$$

where $h_i(s_k)$ are normalised rule firing strengths (non-linear functions of s_k) defined as

$$h_i(s_k) = \frac{\mathcal{T}_{j=1}^{p} \mu_{F_j^i}(s_k^j)}{\sum_{i=1}^{M} (\mathcal{T}_{j=1}^{p} \mu_{F_j^i}(s_k^j))}, \tag{6.10}$$

and $\mathcal{T}$ denotes a t-norm (e.g., a product). The term $\mu_{F_j^i}(s_k^j)$ is the grade of a membership of the premise variable s_k^j. Moreover, the rule firing strengths $h_i(s_k)$ ($i = 1, \ldots, M$) satisfy the following constraints (the convex sum property):

$$\begin{cases} \sum_{i=1}^{M} h_i(s_k) = 1, \\ 0 \leqslant h_i(s_k) \leqslant 1, \qquad \forall i = 1, \ldots, M. \end{cases} \tag{6.11}$$

Hence, (6.9) also corresponds to a quasi-LPV form [14].

6.3 Development of the Takagi–Sugeno Fuzzy Model

Figure 6.1 illustrates the model-based fuzzy control design approach. In order to design a fuzzy controller, a Takagi–Sugeno fuzzy model for a non-linear system is needed. Hence, the construction of a fuzzy model is of a paramount importance and

Fig. 6.1 Model-based fuzzy control design

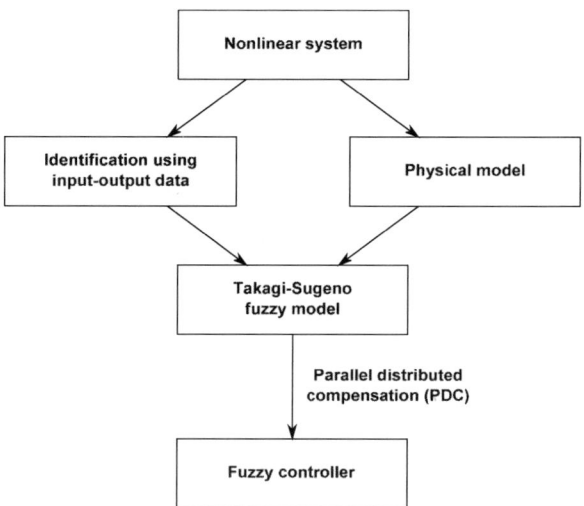

it is an essential step in the approach being considered. As has been mentioned, there are two approaches for constructing fuzzy models:

- identification (fuzzy modelling) using input-output data,
- derivation from given non-linear system equations.

Following the excellent work [15], there is a large set of literature on fuzzy modelling using the input-output data. The procedure essentially consists of two parts: structure identification and parameter identification. The identification approach to fuzzy modelling is suitable for plants for which it is impossible or too difficult to derive an analytical model [7]. A very interesting approach concerning experimental design for the identification of such models is described in [16].

In reality, input-output identification methods allow finding a model in the form (6.7) or an equivalent one using linear/non-linear estimation [12, 13] or clustering methods [17, 18]. The identification problem also demands the selection of premise variables, which govern the changes of a dynamical regime. Practically speaking, to guarantee smooth behaviour of the model, the premise variables are selected to be slowly varying. The combination of local models requires a consistency among them. However, one way to guarantee the consistency of the rules using state-space models is to identify all local models with the same order and convert all of them into the so-called observer canonical form. In this way all the states of the local models will be consistent and in the form of (6.7), whereas their evolution will be perfectly synchronised [17]. On the other hand, non-linear dynamic models for physical systems can be readily obtained by, for example, the Newton–Euler method and the Lagrange method [7]. In such cases, the latter approach, which derives a fuzzy model in the form (6.9) from given non-linear dynamical models, is more appropriate. It utilises the concept of 'sector non-linearity', 'local approximation' or a combination of those to construct fuzzy models. The latter is based on the linearisation around several set points of the non-linear system. In this case, the resulting model is only an approximation and the membership functions $\mu_{F_j^i}(s_k^j)$ are chosen as triangular or sigmoid functions [7], whereas the former technique represents exactly the analysed non-linear model in a compact set of the state variables. In other words, they are composed of linear models blended together with non-linear functions.

6.4 Virtual Fuzzy Actuators

Let us consider the following TS reference model:

$$x_{k+1} = A_k x_k + B_k u_k, \tag{6.12}$$
$$y_{k+1} = C_{k+1} x_{k+1}, \tag{6.13}$$

with $A_k = \sum_{i=1}^{M} h_i(s_k)A^i$, $B_k = \sum_{i=1}^{M} h_i(s_k)B^i$, $C_{k+1} = \sum_{i=1}^{M} h_i(s_{k+1})C^i$ for $i = 1, \ldots, M$. Let us also consider a possibly faulty TS system described by the following equations:

$$x_{f,k+1} = A_k x_{f,k} + B_k u_{f,k} + L_k f_k \qquad (6.14)$$

$$y_{f,k+1} = C_{k+1} x_{f,k+1}, \qquad (6.15)$$

with $L_k = \sum_{i=1}^{M} h_i(s_k)L^i$.

The main objective of this section is to propose a control strategy which can be used for determining the system input $u_{f,k}$ such that

- the control loop for the system (6.14)–(6.15) is stable;
- $x_{f,k+1}$ converges asymptotically to x_{k+1} irrespective of the presence of the fault f_k.

The subsequent part of this section shows development details of the scheme that is able to settle such a challenging problem. The crucial idea is to use the following control strategy:

$$u_{f,k} = -S_k \hat{f}_k + K_{1,k}(x_k - \hat{x}_{f,k}) + u_k. \qquad (6.16)$$

Thus, the following problems arise:

- to determine $\hat{f}_k$;
- to design $K_{1,k}$ in such a way that the control loop is stable, i.e., the stabilisation problem. The control law in such a form is called the PDC, [19],
- to estimate $\hat{x}_{f,k}$.

Let us assume that the following rank condition is satisfied:

$$\text{rank}(C_{k+1}L_k) = \text{rank}(L_k) = s. \qquad (6.17)$$

This implies that it is possible to calculate

$$H_{k+1} = (C_{k+1}L_k)^+ = \left[(C_{k+1}L_k)^T C_{k+1}L_k \right]^{-1} (C_{k+1}L_k)^T. \qquad (6.18)$$

By multiplying (6.15) by H_{k+1} and then substituting (6.14), it can be shown that

$$f_k = H_{k+1}(y_{f,k+1} - C_{k+1}A_k x_{f,k} - C_{k+1}B_k u_{f,k}). \qquad (6.19)$$

Thus, if $\hat{x}_{f,k}$ is used instead of $x_{f,k}$, then the fault estimate is given as follows

$$\hat{f}_k = H_{k+1}(y_{f,k+1} - C_{k+1}A_k \hat{x}_{f,k} - C_{k+1}B_k u_{f,k}) \qquad (6.20)$$

and the associated fault estimation error is

$$f_k - \hat{f}_k = -H_{k+1}C_{k+1}A_k(x_{f,k} - \hat{x}_{f,k}). \tag{6.21}$$

Unfortunately, the crucial problem with practical implementation of (6.20) is that it requires $y_{f,k+1}$ and $u_{f,k}$ to calculate $\hat{f}_k$, and hence it cannot be directly used to obtain (6.16). To settle this problem, it is assumed that there exists a diagonal matrix α_k such that $\hat{f}_k \cong \mathring{f}_k = \alpha_k \hat{f}_{k-1}$, and hence the practical form of (6.16) boils down to

$$u_{f,k} = -S_k \mathring{f}_k + K_{1,k}(x_k - \hat{x}_{f,k}) + u_k. \tag{6.22}$$

In most cases the matrix α_k should be equivalent to an identity one, i.e., it would simply mean a one step delay, which should have a negligible effect on the performance of FTC. In Chap. 5, it was shown how to settle this problem with the $\mathcal{H}_\infty$ approach.

Subsequently, by substituting (6.16) into (6.14), it can be shown that

$$x_{f,k+1} = A_k x_{f,k} - B_k S_k \mathring{f}_k + B_k K_{1,k}(e_k + e_{f,k}) + B_k u_k + L_k f_k, \tag{6.23}$$

where $e_k = x_k - x_{f,k}$ stands for the tracking error while $e_{f,k} = x_{f,k} - \hat{x}_{f,k}$ stands for the state estimation error.

Let us assume that S_k satisfies $B_k S_k = L_k$, e.g., for actuator faults $S_k = I$. Thus

$$x_{f,k+1} = A_k x_{f,k} + L_k(f_k - \mathring{f}_k) + B_k K_{1,k} e_k + B_k K_{1,k} e_{f,k} + B_k u_k. \tag{6.24}$$

Finally, substituting (6.21) into (6.24) and then applying the result into $e_{k+1} = x_{k+1} - x_{f,k+1}$ yield

$$e_{k+1} = (A_k - B_k K_{1,k})e_k + (L_k H_{k+1}C_{k+1}A_k - B_k K_{1,k})e_{f,k}. \tag{6.25}$$

As has already been mentioned, the fault estimate (6.20) is obtained based on the state estimate $\hat{x}_{f,k}$. This raises the necessity for observer design. Consequently, by substituting (6.19) into (6.14), it is possible to show that

$$x_{f,k+1} = \bar{A}_k x_{f,k} + \bar{B}_k u_{f,k} + \bar{L}_k y_{f,k+1}, \tag{6.26}$$

where

$$\bar{A}_k = (I - L_k H_{k+1}C_{k+1})A_k,$$
$$\bar{B}_k = (I - L_k H_{k+1}C_{k+1})B_k, \quad \bar{L}_k = L_k H_{k+1}.$$

Thus, the observer structure, which can be perceived as an UIO (see, e.g., [20, 21]), is given by

$$\hat{x}_{f,k+1/k} = \bar{A}_k \hat{x}_{f,k} + \bar{B}_k u_{f,k} + \bar{L}_k y_{f,k+1},$$
$$\hat{x}_{f,k+1} = \hat{x}_{f,k+1/k} + K_{2,k+1}(y_{f,k+1} - C_{k+1}\hat{x}_{f,k+1/k}). \tag{6.27}$$

Finally, the state estimation error can be written as follows:

$$e_{f,k+1} = (\bar{A}_k - K_{2,k+1}C_{k+1}\bar{A}_k)e_{f,k}. \tag{6.28}$$

The main objective is to summarise the presented results within an integrated framework for the development of the fault identification and fault-tolerant control scheme. First, let us start with two crucial assumptions:

- the pair $(\bar{A}_k, C_{k+1})$ is detectable
- the pair (A_k, B_k) is stabilisable.

Under these assumptions, it is possible to design the matrices $K_{1,k}$ and $K_{2,k}$ in such a way that the extended error

$$\bar{e}_k = \begin{bmatrix} e_k \\ e_{f,k} \end{bmatrix}, \tag{6.29}$$

described by

$$\bar{e}_{k+1} = \begin{bmatrix} A_k - B_k K_{1,k} & L_k H_{k+1} C_{k+1} A_k - B_k K_{1,k} e_{f,k} \\ 0 & \bar{A}_k - K_{2,k+1} C_{k+1} \bar{A}_k \end{bmatrix} \bar{e}_k = A_{e,k} \bar{e}_k, \tag{6.30}$$

converges asymptotically to zero.

It can be observed from the structure of (6.30) that the eigenvalues of the matrix $A_{e,k}$ are the union of those of $A_k - B_k K_{1,k}$ and $\bar{A}_k - K_{2,k+1} C_{k+1} \bar{A}_k$. This clearly indicates that the design of the state feedback and the observer can be carried out independently (the separation principle).

Let us start with controller design with the corresponding tracking error defined by

$$e_{k+1} = [A_k - B_k K_{1,k}]e_k = A_0(h(s_k))e_k, \tag{6.31}$$

where $K_{1,k} = \sum_{i=1}^{M} h_i(s_k)K_1^i$ and the matrix $A_0(h(s_k))$ belongs to a convex polytopic set defined as

$$\mathbb{A}_0 = \left\{ A_0(h(s_k)) : \sum_{i=1}^{M} h_i(s_k) = 1, \ 0 \leqslant h_i(s_k) \leqslant 1 \right.$$

$$A_0(h(s_k)) = \sum_{i=1}^{M}\sum_{j=1}^{M} h_i(s_k)h_j(s_k)A_{0,i,j},$$

$$\left. A_{0,i,j} = \frac{1}{2}(A^i - B^i K_1^j + A^j - B^j K_1^i) \right\}. \tag{6.32}$$

By adapting the general results of the work [22], the following definition is introduced

Definition 6.3 *The tracking error described by (6.31) is robustly convergent to zero in the uncertainty domain (6.32) iff all eigenvalues of $\boldsymbol{A}_0(\boldsymbol{h}(\boldsymbol{s}_k))$ have magnitude less than one for all values of $\boldsymbol{h}(\boldsymbol{s}_k)$ such that $\boldsymbol{A}_0(\boldsymbol{h}(\boldsymbol{s}_k)) \in \mathbb{A}_0$.*

Theorem 6.4 *The tracking error described by (6.31) is robustly convergent to zero in the uncertainty domain (6.32) if there exist matrices $\boldsymbol{Q}_{i,j} \succ \boldsymbol{0}$, $\boldsymbol{G}_1$, $\boldsymbol{W}_j$ such that*

$$\begin{bmatrix} \boldsymbol{G}_1 + \boldsymbol{G}_1^T - \boldsymbol{Q}_{i,j} & * \\ \boldsymbol{N}_{0,i,j} & \boldsymbol{Q}_{m,n} \end{bmatrix} \succ \boldsymbol{0}, \tag{6.33}$$

for all $i, m = 1, \ldots, M$ and $j \geqslant i$, $n \geqslant m$, where $\boldsymbol{N}_{0,i,j} = \frac{1}{2}[(\boldsymbol{A}^i + \boldsymbol{A}^j)\boldsymbol{G}_1 - \boldsymbol{B}^i \boldsymbol{W}_j - \boldsymbol{B}^j \boldsymbol{W}_i]$.

Proof See [22].

Finally, the design procedure boils down to solving the set of $[\frac{1}{2}M(1 + M)]^2$ LMIs (6.33) and then determining $\boldsymbol{K}_1^i = \boldsymbol{W}_i \boldsymbol{G}_1^{-1}$.

Since the controller design procedure is provided, then the observer synthesis framework can be described. To tackle this problem, it is proposed to use a modified version of the celebrated Kalman filter, which can be described as follows:

$$\hat{\boldsymbol{x}}_{f,k+1/k} = \bar{\boldsymbol{A}}_k \hat{\boldsymbol{x}}_{f,k} + \bar{\boldsymbol{B}}_k \boldsymbol{u}_{f,k} + \bar{\boldsymbol{L}}_k \boldsymbol{y}_{f,k+1},$$

$$\boldsymbol{P}_{k+1/k} = \bar{\boldsymbol{A}}_k \boldsymbol{P}_k \bar{\boldsymbol{A}}_k^T + \boldsymbol{U}_k,$$

$$\boldsymbol{K}_{2,k+1} = \boldsymbol{P}_{k+1/k} \boldsymbol{C}_{k+1}^T \left(\boldsymbol{C}_{k+1} \boldsymbol{P}_{k+1/k} \boldsymbol{C}_{k+1}^T + \boldsymbol{V}_{k+1} \right)^{-1},$$

$$\hat{\boldsymbol{x}}_{f,k+1} = \hat{\boldsymbol{x}}_{f,k+1/k} + \boldsymbol{K}_{2,k+1}(\boldsymbol{y}_{f,k+1} - \boldsymbol{C}_{k+1}\hat{\boldsymbol{x}}_{f,k+1/k}),$$

$$\boldsymbol{P}_{k+1} = \left[\boldsymbol{I} - \boldsymbol{K}_{2,k+1}\boldsymbol{C}_{k+1} \right] \boldsymbol{P}_{k+1/k}, \tag{6.34}$$

where $\boldsymbol{U}_k = \delta_1 \boldsymbol{I}$ and $\boldsymbol{V}_k = \delta_2 \boldsymbol{I}$, with δ_1 and δ_2 sufficiently small positive numbers.

It is important to note that the Kalman filter is applied here for state estimation of a deterministic system (6.14)–(6.15), and hence $\boldsymbol{U}_k$ and $\boldsymbol{V}_k$ play the role of the instrumental matrices only (see Sect. 2.3 for more details).

When the initial tracking error is known (i.e., the deviation of a faulty system state from a nominal system state), an upper bound on the norm of the control input $\mathring{\boldsymbol{u}}_{f,k} = \boldsymbol{K}_{1,k}(\boldsymbol{x}_k - \hat{\boldsymbol{x}}_{f,k})$ can be found as follows [23]: Let us assume that the initial tracking error $\boldsymbol{e}_0$ lies in an ellipsoid of the diameter γ, i.e., $\|\boldsymbol{e}_0\| \leqslant \gamma$, then the constraint on a control input described as follows $\|\mathring{\boldsymbol{u}}_{f,k}\|_{\max} = \max_l \left| \mathring{\boldsymbol{u}}_{f,k}^l \right| \leq \lambda$ is enforced at all times if the following LMIs hold:

$$\begin{bmatrix} \boldsymbol{X} & * \\ 0.5(\boldsymbol{W}_i^T + \boldsymbol{W}_j^T)\,\boldsymbol{G}_1 + \boldsymbol{G}_1^T - \boldsymbol{Q}_{i,j} \end{bmatrix} \succeq \boldsymbol{0},$$

$$\boldsymbol{Q}_{i,j} \succeq \gamma^2 \boldsymbol{I}, \quad \mathrm{diag}(\boldsymbol{X}) \preceq \lambda^2 \boldsymbol{I}, \tag{6.35}$$

where W_i, W_j, $Q_{i,j}$ and G_1 satisfy conditions given by (6.33) for all $i = 1, \ldots, M$ and $j \geqslant i$.

In order to solve the regulator problem it is necessary to find a state feedback controller such that the following objective function is minimised:

$$J_\infty = \sum_{i=0}^{\infty} \left(\mathring{\boldsymbol{y}}_{f,k+i}^T \boldsymbol{Q}_R \mathring{\boldsymbol{y}}_{f,k+i} + \mathring{\boldsymbol{u}}_{f,k+i}^T \boldsymbol{R}_R \mathring{\boldsymbol{u}}_{f,k+i} \right), \qquad (6.36)$$

where $\mathring{\boldsymbol{y}}_{f,k+i} = \boldsymbol{y}_{f,k+i} - \boldsymbol{y}_{k+i}$, $\boldsymbol{Q}_R \succeq 0$ and $\boldsymbol{R}_R \succ 0$ are suitable weight matrices.

However, the system being considered is uncertain, and hence, only the upper bound of the objective function can be minimised. Therefore the following theorem gives only a sub-optimal solution for the regulator problem [22].

Theorem 6.5 *The upper bound for the objective function* (6.36) *for the initial tracking error* $\boldsymbol{e}_0$ *lying in an ellipsoid of the diameter* γ *can be obtained by solving the following LMI optimization problem:*

$$\min_{\boldsymbol{Q}_{i,j},\boldsymbol{G}_1,\boldsymbol{W}_i} \eta,$$

subject to

$$\begin{bmatrix} \boldsymbol{G}_1 + \boldsymbol{G}_1^T - \boldsymbol{Q}_{i,j} & * & * & * \\ \boldsymbol{N}_{0,i,j} & \boldsymbol{Q}_{m,n} & * & * \\ 0.5\boldsymbol{Q}_R^{1/2}(\boldsymbol{C}^i + \boldsymbol{C}^j)\boldsymbol{G}_1 & 0 & \eta\boldsymbol{I} & * \\ 0.5\boldsymbol{R}_R^{1/2}(\boldsymbol{W}_i + \boldsymbol{W}_j) & 0 & 0 & \eta\boldsymbol{I} \end{bmatrix} \succ \boldsymbol{0}, \quad \boldsymbol{Q}_{i,j} \succeq \gamma^2\boldsymbol{I}, \qquad (6.37)$$

for all $i, m = 1, \ldots, M$ *and* $j \geqslant i, n \geqslant m$, *where* $\boldsymbol{N}_{0,i,j} = \frac{1}{2}[(\boldsymbol{A}^i + \boldsymbol{A}^j)\boldsymbol{G}_1 - \boldsymbol{B}^i \boldsymbol{W}_j - \boldsymbol{B}^j \boldsymbol{W}_i]$ *and the local feedback gains are* $\boldsymbol{K}_1^i = \boldsymbol{W}_i \boldsymbol{G}_1^{-1}$.

Proof See [22].

6.4.1 Implementation Details

This section provides a clear design procedure regarding the proposed approach. The initial stage is to compute virtual actuator gains $\boldsymbol{K}_1^i$ for all $i = 1, \ldots, M$, by solving LMIs described by (6.33) or (6.37) (if a regulator is needed); the specific input constraints (6.35) can be considered if required. Finally, a virtual actuator is developed, which is shown in Fig. 6.2.

The first step is to compute the virtual actuator output described by (6.22). To do so, there is a need for using the current control input, the difference between a previously estimated state of a possibly faulty system and a state of a nominal system. Because the current fault estimate is not available, there is a need for using

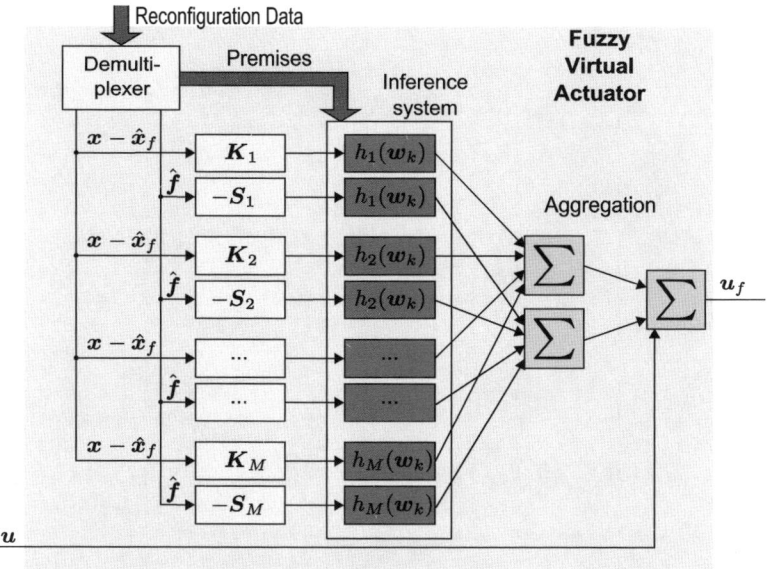

Fig. 6.2 Virtual actuator for Takagi–Sugeno fuzzy systems

$\overset{\circ}{f}_k = \alpha_k \hat{f}_{k-1}$. Finally, it is feasible to find an output of a possibly faulty system achieved by using the new control input (from the virtual actuator) and also to find the reference system state using the nominal input (6.12). Indeed, it is possible to compute an estimate of the state of a possibly faulty system by using (6.27), and hence to compute compute the fault estimate by using (6.19).

6.5 Virtual Fuzzy Sensor

One observer is sufficient to detect faults, i.e., it can indicate an alarm if a fault occurs in the system. To isolate faults, a number of observers should be designed based on the idea of *generalised observer scheme* [24, 25] and has already been successfully used in FTC systems [26, 27]. The generalised observer scheme used as a virtual sensor is depicted in Fig. 6.3. The upper block in a fuzzy virtual sensor is the observer (integrated with the model) of the overall system. By means of its output it checks the consistency of each of the input-output pairs. If $\varepsilon(k)$ exceeds some predetermined threshold T^S (determined by the noise and modelling inaccuracy of the original system), which may not be constant (the adaptive threshold is favoured, i.e., a threshold which is changing according to the perceived accuracy or robustness of the model at any given time point), then the outputs of other m observers must be checked. The fault detection and isolation task can be efficiently solved with the concept of a TS virtual sensor. Each of the m blocks is fed by all but one output signal and tests the consistency of the estimated ith output and its sensor readings. If the resulting residual $z_{i,k}$ does not exceed a predefined threshold T_i^S, then this means

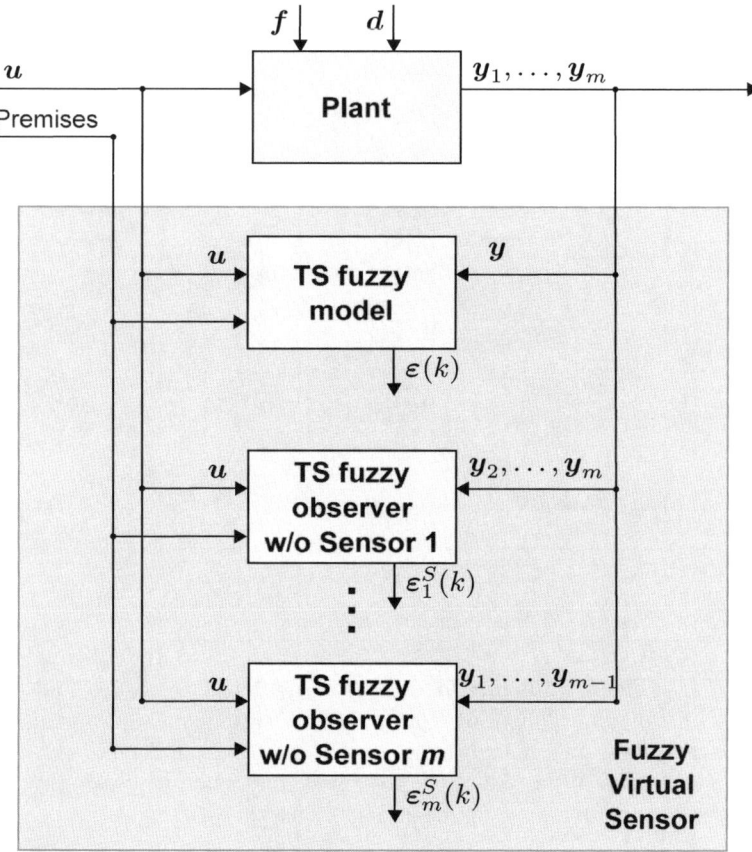

Fig. 6.3 Virtual sensor for Takagi–Sugeno fuzzy systems

that the ith sensor performs correctly. After a sensor fault is detected and isolated, the FTC system is reconfigured to use all but the ith corrupted sensor readings, while the output of the ith virtual sensor replaces the output of the faulty sensor. After the sensor is repaired (or replaced), the use of all sensors can be safely resumed. Finally, it should be pointed out that the approach presented in this section is very similar to that of Sect. 5.3.2.

6.6 Illustrative Examples

The main objective of the subsequent part of this section is to provide illustrative examples that will exhibit the performance of the approaches described in the present chapter. In particular, the systems being considered include: a tunnel furnace, a three-tank system and a twin-rotor system. All examples present a comprehensive case study regarding the selected approaches.

6.6.1 TS Model Design for a Tunnel Furnace

The tunnel furnace is a laboratory counterpart of real industrial tunnel furnaces, which can be applied in the food industry or production of ceramics (among others). The furnace is equipped with three electric heaters and four temperature sensors. The maximum power outputs of the heaters were measured to be approximately 686 W, 693 W and 756 W ± 20 W, respectively. The required temperature of the furnace can be kept by controlling the work of the heaters. This task can be achieved by group regulation of the voltage with the application of the controller PACSystems RX3i (with the firmware version 6.0) manufactured by GE Fanuc Intelligent Platforms [28] and the semiconductor relays RP6 produced by LUMEL [29], providing an impulse control with a variable impulse frequency $f_{max} = 1$ Hz. The temperature of the furnace is measured via IC695ALG600 module [30] from Pt100 Resistive Thermal Devices (RTDs) with the accuracy of $\pm 0.7\,°C$. The visualisation of the work of the tunnel furnace is realised with the Quickpanel View device from GE Fanuc Intelligent Platforms [31]. Its hardware setup can be seen in Figs. 6.4 and 6.5. It is worth noting that the system considered is a distributed–parameter one (i.e., a system whose state space is infinite-dimensional), thus any resulting model from input-output data will be at best an approximation. Hence, to achieve a good approximation, the optimal experimental design is of paramount importance and is often an iterative process. An interesting paper dealing specifically with the identification of the TS fuzzy models is [16], and its recommendations are partially repeated in a book dealing with fuzzy systems [17].

In non-linear system identification, both the amplitude and frequency contents of the input signals are of major importance. Thus, for identifying TS fuzzy models containing both equilibrium and off-equilibrium local affine models, it is recommended that input signals should be designed according to the following guidelines [16]:

- The system being considered should be brought through a sequence of equilibria that includes the equilibria of the local models. At each equilibrium the system should be excited by super-positioned PRBS, i.e., a pseudo–random, usually binary signal added to the original input signal. The PRBS signals should have the frequency content that covers an interval from the inverse rise time to above the bandwidth of the closed-loop system.
- For each off-equilibrium local model, several transient trajectories should be generated. The corresponding input signals should contain both large amplitude steps and perturbations, so both the trend and perturbation dynamics of the off-equilibrium local models could be determined. Also the frequency contents should typically be higher compared to the frequency content of the equilibrium data to prevent the system from settling at some equilibrium.

Of course, these are general guidelines, so in practical applications there will be some constraints that will often limit the number of transitions, frequency content, amplitudes, and the length of the experiment. Also depending strongly on the application, the requirements in terms of the accuracy of off-equilibrium local models

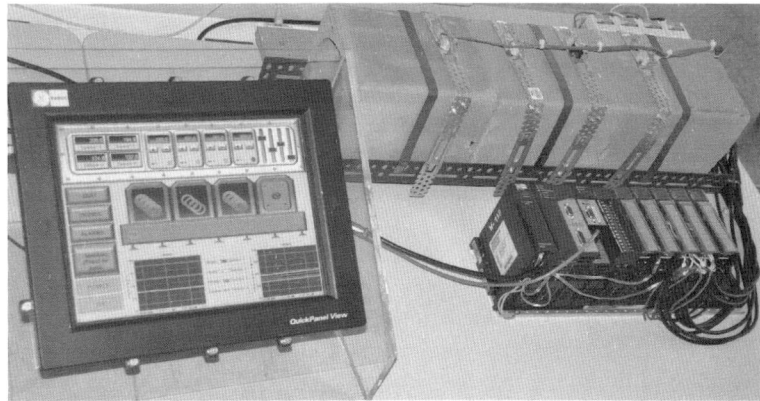

Fig. 6.4 Laboratory model of a tunnel furnace—hardware setup

Fig. 6.5 Interior of a tunnel furnace

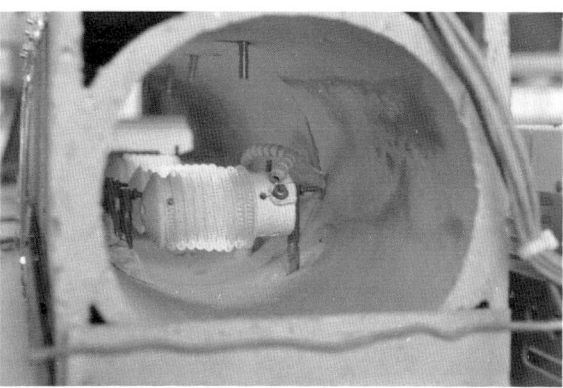

should be considered. Sometimes, equilibrium local models can be extrapolated into transient operating regions without significant loss of accuracy.

Other excitation signals are multisine signals with variable frequency and the swept sinus with random frequencies. These signals are frequently used in the identification of mechanical systems [17].

Thus in order to identify a model for the tunnel furnace, the input signals were defined as follows:

- Five operating points were considered, at 20, 40, 60, 80 and 100 % of the maximum power of heaters, respectively.
- *Heating phase*: at each operating point for the first 1,800 s, the constant input signal values were used (to heat the furnace to a desired temperature).
- *Perturbation phase*: after the heating phase, for the next 1,620 s (for each of the input signal individually and independently) a perturbation signal was applied as

follows: the perturbation was a uniformly distributed pseudo-random signal in the range of $[-10, 10\%]$ of the maximum power, super-positioned on the signal generated in the heating phase (for 100% operating point the range was $[-10, 0\%]$), whereas each signal duration was chosen at random in a range of [5,15] seconds. Finally, a new value and duration were generated.

- *Short cooling phase*: after the perturbation phase, for a short time period of 120 s, the heaters were disabled.
- *Short heating phase*: after the cooling phase, for 60 s, a maximum power for all the heaters was applied.
- Thus, a cycle for each operating point lasted 3,600 s, and after five full cycles, for the remaining time of simulation, a uniformly distributed pseudo-random signal in the range of $[0, 100\%]$ of maximum power and duration in the range of [5,15] seconds were applied for each heater individually, thus, giving a total duration of the experiment equal to 19,000 s.

Afterwards, the resulting experimental data were cut into appropriate segments for each of the operating points. From each input-output data of the operating point, a local system-state model was built, using subspace methods. Subspace methods originate in a mix between system theory, geometry and numerical linear algebra. These subspace methods are successfully used for model identification for industrial processes [32]. The N4SID [33] algorithm was selected for the task of model identification, with the order of the models equal to four, as greater values did not introduce a significant increase in the modelling accuracy. Subsequently, to guarantee the consistency of fuzzy rules, all of the resulting local models were converted into the observer canonical form [33]. The resulting matrices for each local model are

$$
A^1 = \begin{bmatrix} 1.0021 & -0.0040 & -0.0230 & 0.0259 \\ 0.0023 & 0.9960 & -0.0083 & 0.0099 \\ 0.0024 & -0.0028 & 0.9907 & 0.0099 \\ 0.0009 & -0.0005 & -0.0059 & 1.0051 \end{bmatrix},
$$

$$
A^2 = \begin{bmatrix} 0.9995 & -0.0048 & 0.0010 & 0.0038 \\ 0.0003 & 0.9956 & 0.0008 & 0.0028 \\ 0.0001 & -0.0014 & 0.9994 & 0.0011 \\ 0.0002 & -0.0023 & 0.0005 & 1.0011 \end{bmatrix},
$$

$$
A^3 = \begin{bmatrix} 1.0013 & -0.0034 & 0.0024 & -0.0009 \\ 0.0021 & 0.9970 & 0.0002 & 0.0004 \\ 0.0011 & -0.0002 & 0.9976 & 0.0013 \\ 0.0006 & 0.0001 & -0.0006 & 0.9993 \end{bmatrix},
$$

$$A^4 = \begin{bmatrix} 0.9993 & -0.0022 & 0 & 0.0029 \\ 0.0002 & 0.9967 & -0.0005 & 0.0034 \\ -0.0003 & 0.0001 & 0.9987 & 0.0006 \\ 0.0004 & -0.0018 & 0.0016 & 0.9989 \end{bmatrix},$$

$$A^5 = \begin{bmatrix} 0.9977 & -0.0054 & 0.0065 & 0.0002 \\ 0.0003 & 0.9925 & 0.0063 & 0 \\ -0.0030 & -0.0025 & 1.0071 & -0.0035 \\ -0.0046 & -0.0011 & 0.0093 & 0.9944 \end{bmatrix},$$

$$B^1 = \begin{bmatrix} 0.4565 & 0.7132 & -0.3372 \\ 0.2529 & 0.5025 & -0.1768 \\ -0.0991 & 0.2829 & 0.2101 \\ 0.0196 & 0.1951 & 0.1526 \end{bmatrix},$$

$$B^2 = \begin{bmatrix} 0.0590 & -0.0169 & 0.5376 \\ 0.1541 & 0.1590 & 0.2107 \\ 0.0760 & 0.0264 & 0.3059 \\ -0.1752 & 0.1992 & 0.3504 \end{bmatrix},$$

$$B^3 = \begin{bmatrix} 0.1647 & 0.1054 & -0.0362 \\ 0.0853 & 0.1502 & -0.0291 \\ 0.0295 & 0.1020 & 0.0665 \\ 0.0184 & 0.0513 & 0.1089 \end{bmatrix},$$

$$B^4 = \begin{bmatrix} 0.0090 & 0.3650 & -0.0617 \\ 0.2099 & 0.4394 & -0.2910 \\ 0.2848 & 0.3058 & -0.2542 \\ 0.2192 & 0.2708 & -0.2352 \end{bmatrix},$$

$$B^5 = \begin{bmatrix} 0.0383 & -0.0081 & 0.3631 \\ 0.3795 & -0.2730 & 0.2910 \\ 0.4411 & -0.4493 & 0.3601 \\ 0.7984 & -0.4391 & -0.1311 \end{bmatrix}.$$

Finally, as the premise variable, the second temperature output was chosen with triangular membership functions. To improve the approximation results, the parameters of the membership functions were optimised. The algorithm used for this task

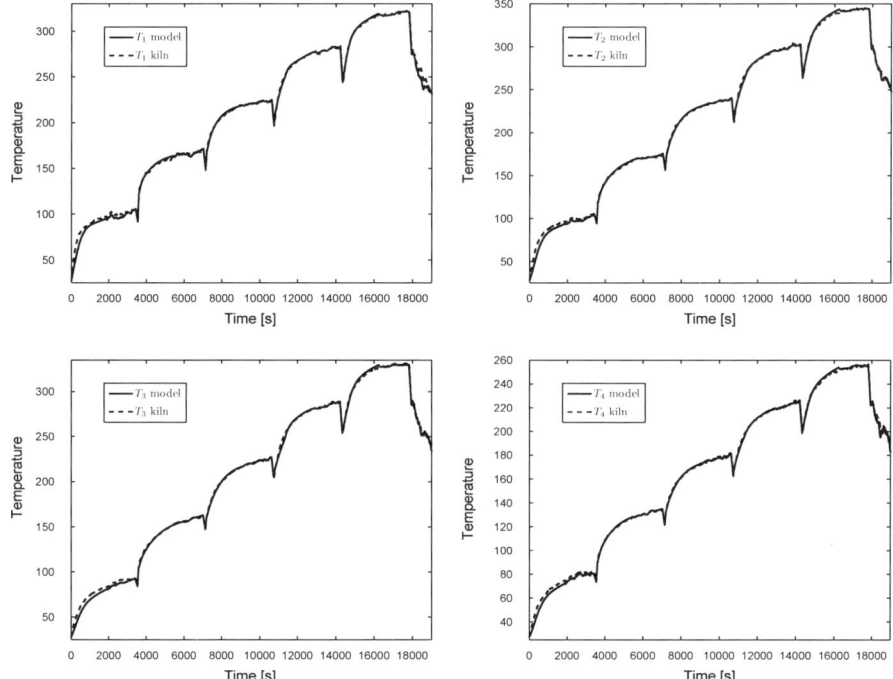

Fig. 6.6 Response of the Takagi–Sugeno model and the tunnel furnace

was a non-linear least-squares solver implemented in MATLAB environment, which utilised the trust-region-reflective algorithm. This algorithm is a subspace trust-region method and is based on the interior-reflective Newton method described in [34]. The modelling results can be seen in Fig. 6.6, whilst the control trajectories and fuzzy sets are shown in Fig. 6.7. Each figure presents the response of the tunnel furnace and the TS model with the fuzzy membership functions shown in Fig. 6.7. As can be observed, the resulting model is quite good and follows the original trajectories with only a slight error, which can be observed mostly for the first TS model.

6.6.2 Virtual Sensor for a Tunnel Furnace

The definition of reference inputs is exactly the same as that in the preceding section, the only difference is that the first hour of the experiment is considered (for the clarity of presentation). The four sensors are available and normally provide real experimental data with an additional assumption that initially all sensors are healthy (any discrepancy at this point would be clearly visible to anyone). The following sensor fault scenario is considered here:

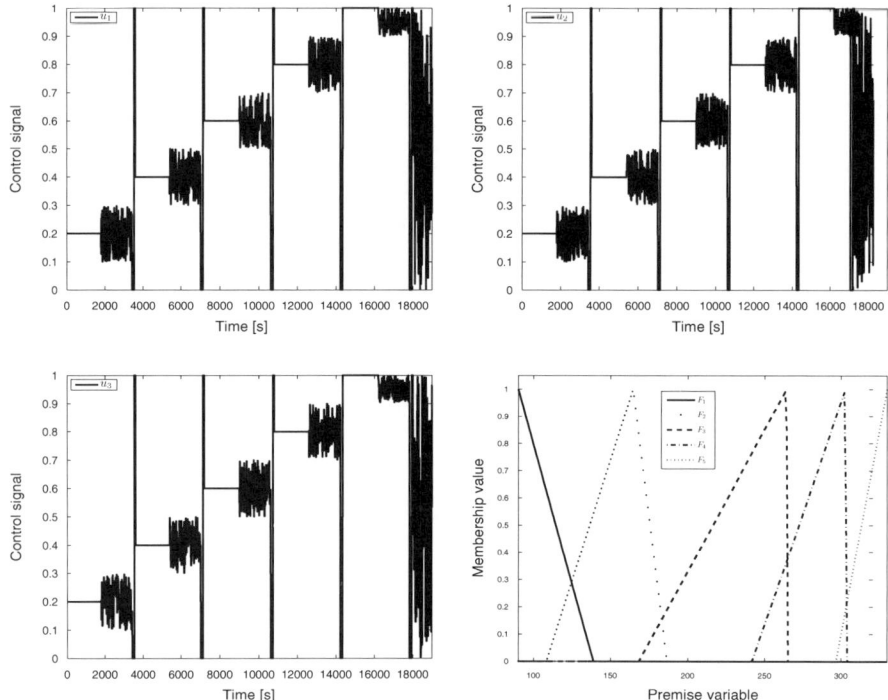

Fig. 6.7 Selected input trajectories and fuzzy sets used in the Takagi–Sugeno model

- The first sensor's total failure, i.e., it is switching to constant reading of 20 °C or can be considered disconnected. The fault is assumed to start at the 2,000th s and lasts till 2,500th s, when it is assumed to be repaired (replaced).
- The third sensor being stuck, i.e., it is switching to reading its last healthy measurement. The fault is assumed to start at the 500th s and lasts till 1,500th when it is assumed to be repaired (replaced or restarted).
- The rest of the sensors are assumed to be healthy.

The results of experiment can be seen in Figs. 6.8 and 6.9. In Fig. 6.8, residuals provided by four virtual sensors are provided. Figure 6.8 portrays the performance of the virtual sensors. It can be easily seen that the residuals indicate the occurrence of faults. It can also be observed that the estimates of the signals measured by the faulty sensors are correct. Let us start the analysis with the first sensor fault. This kind of abrupt fault, being the total failure (or disconnection, etc.) of the sensor, is the most common one. In fact, the hardware driving the tunnel furnace is capable of detecting such errors by itself and allowing the repairs to be made on-line. But assuming that the fault was not detected by the hardware, for example due to some kind of short-circuit (but with the magnitude of several Ohms) in the wiring of the sensor, it would still be detected by the use of the virtual sensors. By observing

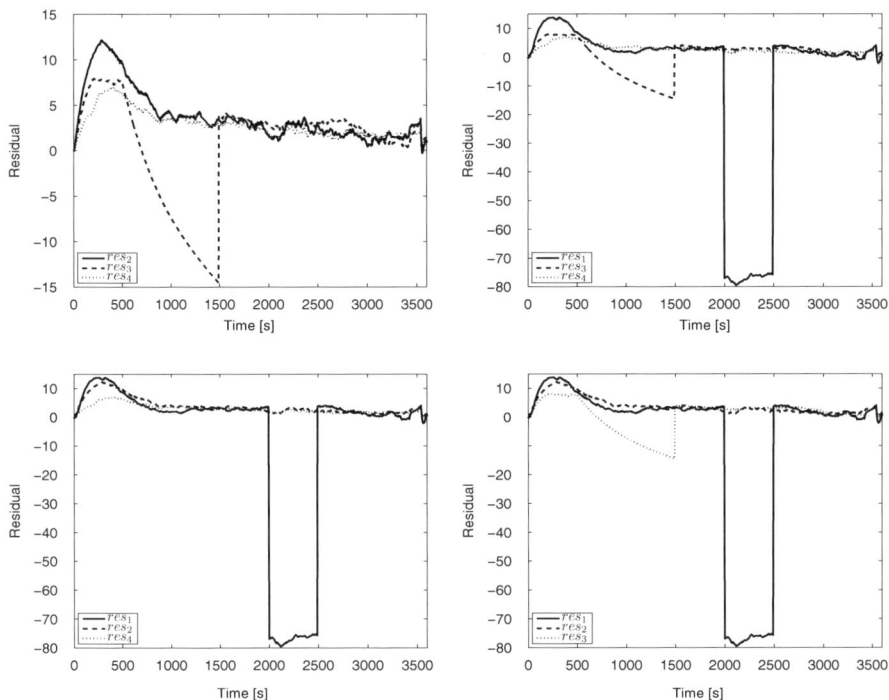

Fig. 6.8 Residuals obtained with the virtual sensors

residuals between the 2,000th and the 2,500th s, it can be seen that all but first sensor residuals show an abrupt change, whereas the first virtual sensor residuals show only the standard low level variations in the residuals. Thus, this fault can be easily isolated and its effect can be neglected by substituting the faulty reading with the output of the corresponding virtual sensor. After the sensor is repaired, residual drops in all virtual sensors and readings from all the outputs can be safely resumed.

The next fault being considered concerns the third sensor. Such a fault would be very uncommon in the hardware configuration of the tunnel furnace, considering that a temperature sensor is an electrical one, but can be a common one in mechanical sensors. Indeed, most critical safety industrial applications (power plants, central heating, etc.) would require the use of mechanical indicators and the electrical sensors in a tandem, providing redundancy and supervision by a human operator, who would make a similar comparison of readings as the one considered here.

6.6.3 FTC of a Three-Tank System

The selected non-linear system results from a celebrated benchmark [35] and is shown in Fig. 6.11. It portrays a hydraulic process consisting of three identical tanks

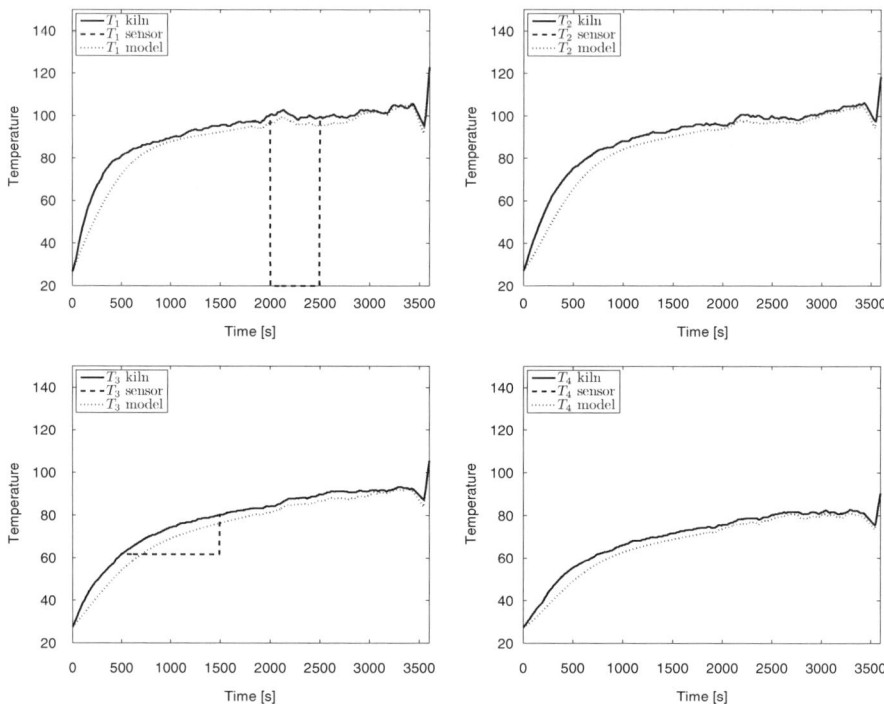

Fig. 6.9 Performance of virtual sensors

T_1, T_2, T_3 with a cross section A. These tanks are connected to each other by cylin-drical pipes with identical cross sections S_n. The nominal outflow valve is located at the outlet of the tank T_2. The out-flowing liquid is collected in a reservoir (of a greater volume than all tanks combined), which supplies the pumps 1 and 2. Q_1 and Q_2 are the flow rates of the pumps 1 and 2, respectively. The water lev-els x_1 and x_2 are measured via piezoresistive pressure sensors. The connecting pipes between the tanks are equipped with manually adjustable ball valves, which allow the corresponding pipes to be opened or closed in order to simulate clogging or operating errors.

By taking into account the fundamental laws of conservation of fluid, a detailed mathematical model describing dynamic behaviour of this multi-input/multi-output system can be developed. The water levels x_1, x_2 and x_3 are governed by the con-straint $x_1 > x_3 > x_2$, and a non-linear system model is expressed by the following state equations [35]:

$$
\begin{cases}
A\dfrac{dx_1(t)}{dt} = Q_1(t) - \alpha_1 S_n(2g(x_1(t) - x_3(t)))^{1/2} \\
\qquad\qquad + Qf_1(t), \\[4pt]
A\dfrac{dx_2(t)}{dt} = Q_2(t) + \alpha_3 S_n(2g(x_3(t) - x_2(t)))^{1/2} \\
\qquad\qquad - \alpha_2 S_n(2gx_2(t))^{1/2} + Qf_2(t), \\[4pt]
A\dfrac{dx_3(t)}{dt} = \alpha_1 S_n(2g(x_1(t) - x_3(t)))^{1/2} \\
\qquad\qquad - \alpha_3 S_n(2g(x_3(t) - x_2(t)))^{1/2} + Qf_3(t),
\end{cases}
\tag{6.38}
$$

where α_i, $i = 1, 2, 3$, are constants. $Qf_i(t)$, $i = 1, 2, 3$, denote an additional mass flow caused by leaks or actuator faults and g is the gravity constant.

A normalised TS model, with $s(t) = u(t)$, which approximates the non-linear system (6.38), is described in [36]:

$$
\dot{x}(t) = \sum_{i=1}^{4} h_i(u(t))[A^i x(t) + Bu(t) + Lf(t) + d^i],
\tag{6.39}
$$

$$
y(t) = Cx(t).
\tag{6.40}
$$

The matrices A^i, B and d (trend) are acquired by linearising the initial system (6.38) around four points chosen in the operating range of the system considered and the membership functions are shown in Fig. 6.10. Four local models (in this case) guarantee a good approximation of the non-linear state space model of the real system by the TS model. The following numerical values were used:

$$
A^1 = \begin{bmatrix} -0.0109 & 0 & 0.0109 \\ 0 & -0.0206 & 0.0106 \\ 0.0109 & 0.0106 & -0.0215 \end{bmatrix}, \; d^1 = 10^{-3} \begin{bmatrix} -2.86 \\ -0.38 \\ 0.11 \end{bmatrix},
$$

$$
A^2 = \begin{bmatrix} -0.0110 & 0 & 0.0110 \\ 0 & -0.0205 & 0.01044 \\ 0.0110 & 0.01044 & -0.0215 \end{bmatrix}, \; d^2 = 10^{-3} \begin{bmatrix} -2.86 \\ -0.34 \\ 0.038 \end{bmatrix},
$$

$$
A^3 = \begin{bmatrix} -0.0084 & 0 & 0.0084 \\ 0 & -0.0206 & 0.0095 \\ 0.0084 & 0.0095 & -0.0180 \end{bmatrix}, \; d^3 = 10^{-3} \begin{bmatrix} -3.70 \\ -0.14 \\ 0.69 \end{bmatrix},
$$

$$
A^4 = \begin{bmatrix} -0.0085 & 0 & 0.0085 \\ 0 & -0.0205 & 0.0095 \\ 0.0085 & 0.0095 & -0.0180 \end{bmatrix}, \; d^4 = 10^{-3} \begin{bmatrix} -3.67 \\ -0.18 \\ 0.62 \end{bmatrix},
$$

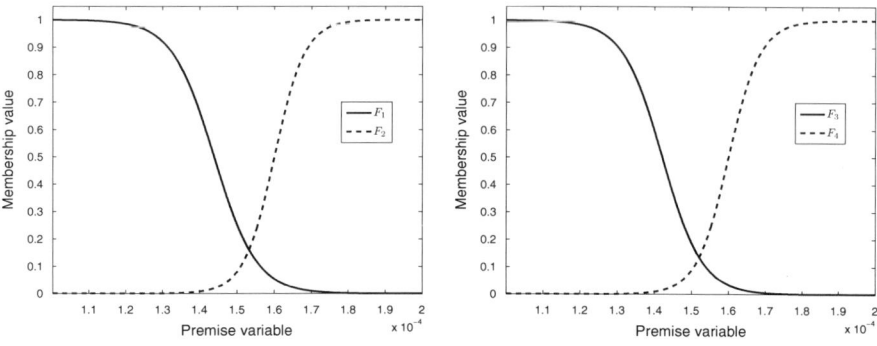

Fig. 6.10 Fuzzy sets used in Takagi–Sugeno model of a three-tank system

$$\boldsymbol{B} = \frac{1}{A}\begin{bmatrix} 1 & 0 \\ 0 & 1 \\ 0 & 0 \end{bmatrix}, \quad \boldsymbol{C} = \begin{bmatrix} 1 & 0 & 0 \\ 0 & 1 & 0 \end{bmatrix}, \quad \boldsymbol{L} = -\boldsymbol{B},$$

while the numerical evaluation of (6.38) was performed with

$$\alpha_1 = 0.78, \quad \alpha_2 = 0.78, \quad \alpha_3 = 0.75,$$
$$g = 9.8, \ S_n = 5 \times 10^{-5} \text{ and } A = 0.0154.$$

Subsequently, the continuous time model (6.39) was transformed into the discrete-time one using the Euler method with the sampling time 100 ms. The reference input is defined by

$$\begin{aligned}
\boldsymbol{u}_{k,1} &= 2.5 \cdot 10^{-5}[\sin(k/320) + 0.5\sin(k/160) \\
&\quad + 0.5\sin(k/640) + 0.3\sin(k/60)] + 10^{-4}, \\
\boldsymbol{u}_{k,2} &= 3.75 \cdot 10^{-6}[\cos(k/320) + 0.5\sin(k/160) \\
&\quad + 0.5\cos(k/640) + 0.3\sin(k/60)] + 1.75 \cdot 10^{-5}.
\end{aligned}$$

The fault scenario is described as follows:

$$\boldsymbol{f}_k = \begin{bmatrix} r_{k,1} & 0 \\ 0 & r_{k,2} \end{bmatrix} \boldsymbol{u}_k,$$

where

$$r_{k,1} = \begin{cases} 0, & k < 1500, \\ 0.4, & k \geqslant 1500, \end{cases}$$

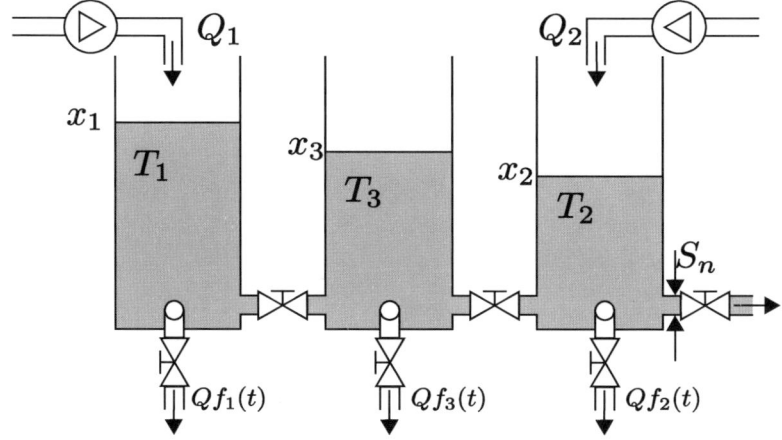

Fig. 6.11 Three-tank system

$$r_{k,2} = \begin{cases} 0, & k < 1000, \\ 0.5 + 0.3\sin(\pi k/90), & k \geqslant 1000. \end{cases}$$

Figure 6.12 presents the achieved results for the proposed FTC strategy (with $\alpha_k = 1$). As a result, Fig. 6.12 clearly shows that the faults can be estimated with very high accuracy. The initial discrepancies are caused by the differences between the initial state of the system and the FTC observer as well as by some model errors. From Fig. 6.13, it can be observed that $u_{f,k,1}$ is equal to $u_{k,1}$ until the occurrence of the fault $f_{k,1}$. After that time the control strategy $u_{f,k,1}$ was changed. It was also changed when f_2 occurred, but due to reasons stated above the FTC system at the beginning tried to achieve an equilibrium ($u_{f,k,2}$ converged to $u_{k,2}$). The final conclusion is that the residual (Fig. 6.14) is very close to zero in the presence of faults.

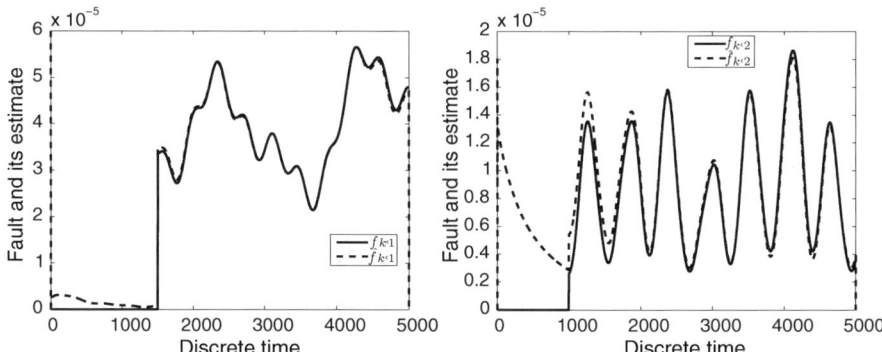

Fig. 6.12 Faults and their estimates

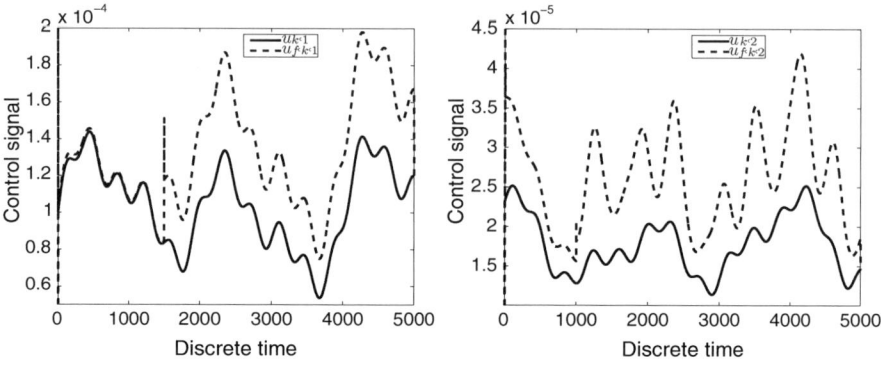

Fig. 6.13 Trajectories of u_k and $u_{f,k}$

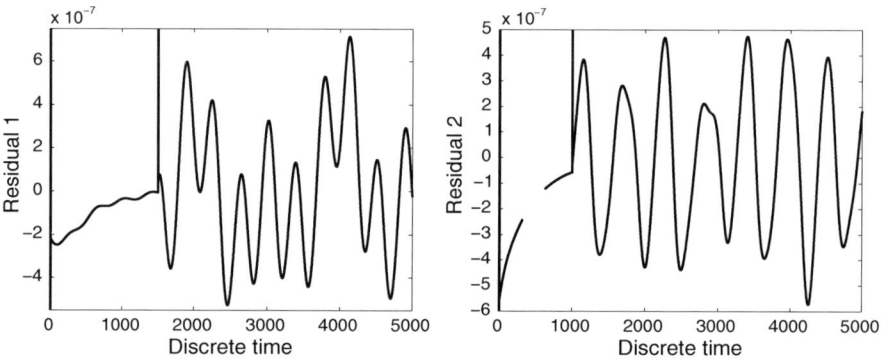

Fig. 6.14 Residual z_k

This is because of the proposed control strategy, for which $x_{f,k}$ converges to x_k and consequently z_k converges to zero. On the other hand, the presence of faults can be easily determined using the fault estimate.

6.6.4 FTC of a Twin-Rotor System

The selected non-linear system is based on the TRMS, a laboratory set-up developed by Feedback Instruments Limited [37] for control experiments. Due to its high non-linearity, cross coupling between its two axes and inaccessibility of some its outputs and states for measurements, the system is often perceived as a challenging engineering problem. Extensive research on the modelling of such a system can be found in [38] and the references therein. The TRMS as shown in Fig. 4.2 is driven by two DC motors. It has two propellers perpendicular to each other and joined by a beam pivoted on its base, so that it can rotate in such a way that its ends move on spherical surfaces. The joined beam can be moved by changing the input voltage of its motor, which controls the rotational speed of the propellers. The system is equipped with a pendulum counterweight fixed to the beam and it determines a stable equilibrium position. Additionally, the system is balanced in such a way that when the motors are switched off, the main rotor end of the beam is lowered. In certain aspects the behaviour of the TRMS system resembles that of a helicopter [38]. For example, there is a strong cross-coupling between the main rotor (collective) and the tail rotor. However, the system is different from a helicopter in many ways, the main differences being the: location of the pivot point (midway between two rotors in TRMS versus main rotor head in the helicopter), vertical control (speed control of main rotor versus collective pitch control), yaw control (tail rotor speed versus pitch angle of tail rotor blades) and lastly, cyclical control (none versus directional control).

The mathematical model of the TRMS can be described by a set of four nonlinear differential equations with two linear differential equations and four non-linear functions [37]. Some of the parameters can be obtained from [37], whereas others should be obtained by experimentations, e.g., inertia, magnitudes of the physical propeller, coefficients of friction and impulse force. The inputs of the system are defined by the input vector $u = [u_h, u_v]^T$, where u_h is the input voltage of the tail motor and u_v is the input voltage of the main motor. The state vector is defined as $x = [\Omega_h, \alpha_h, \omega_t, \Omega_v, \alpha_v, \omega_m]^T$, where Ω_h is the angular velocity around the vertical axis, α_h is the azimuth angle of the beam, ω_t is the rotational velocity of the tail rotor, Ω_v is the angular velocity around the horizontal axis, α_v is the pitch angle of the beam, ω_m is the rotational velocity of the main rotor. The output vector is defined as $y = [\omega_m, \alpha_v, \alpha_h]^T$. For the complete physical model of such a system refer to [37, 38].

A normalised TS model, which approximates the non-linear TRMS system, is obtained by linearising a system around five operating points [39]. The system can be described in the following way:

$$x_{k+1} = \sum_{i=1}^{5} h_i(\alpha_{h,k})[A^i x_k + B^i (u_k - u^i) + L^i f_k], \qquad (6.41)$$

$$y_k = C^i x_k + d^i. \qquad (6.42)$$

The matrices A^i, B^i, C^i, u^i and d^i (trend) are acquired by linearising the initial system around five points chosen in the operating range of the system considered, with the premise variable $s_k = \alpha_{h,k}$ and membership functions as shown in Fig. 6.18. Five local models guarantee a good approximation of the state of the real system by the TS model within the operating range. The following numerical values, with the sampling time 50 ms, were used:

$$A^1 = \begin{bmatrix} 0.9812 & -0.0105 & 0.1847 & 0 & 0 & 0 \\ 0 & 0.9657 & 0 & 0 & 0 & 0 \\ 0 & 0 & 0.8780 & 0 & 0 & 0 \\ 0 & 0.0152 & -0.0254 & 0.9908 & -0.1718 & 0 \\ 0 & 0.0004 & 0.1367 & 0.0498 & 0.9957 & 0 \\ 0.0495 & 0.0276 & 0.0047 & 0 & 0 & 1 \end{bmatrix}, \quad d^1 = \begin{bmatrix} 0 \\ -0.9326 \\ 0 \end{bmatrix},$$

$$A^2 = \begin{bmatrix} 0.9814 & -0.0103 & 0.1841 & 0 & 0.0004 & 0 \\ 0 & 0.9657 & 0 & 0 & 0 & 0 \\ 0 & 0 & 0.8780 & 0 & 0 & 0 \\ 0 & 0.0200 & -0.0254 & 0.9908 & -0.1718 & 0 \\ 0 & 0.0005 & 0.1367 & 0.0498 & 0.9957 & 0 \\ 0.0495 & 0.0274 & 0.0046 & 0 & -0.0010 & 1 \end{bmatrix}, \quad d^2 = \begin{bmatrix} 0.1074 \\ -0.9257 \\ 64.1737 \end{bmatrix},$$

$$A^3 = \begin{bmatrix} 0.9818 & -0.0098 & 0.1830 & 0 & 0.0007 & 0 \\ 0 & 0.9657 & 0 & 0 & 0 & 0 \\ 0 & 0 & 0.8780 & 0 & 0 & 0 \\ 0 & 0.0405 & -0.0254 & 0.9908 & -0.1718 & 0 \\ 0 & 0.0010 & 0.1367 & 0.0498 & 0.9957 & 0 \\ 0.0495 & 0.0268 & 0.0045 & -0.0001 & -0.0020 & 1 \end{bmatrix}, \quad d^3 = \begin{bmatrix} 0.2146 \\ -0.9133 \\ 127.7300 \end{bmatrix},$$

$$A^4 = \begin{bmatrix} 0.9826 & -0.0090 & 0.1809 & 0 & 0.0010 & 0 \\ 0 & 0.9657 & 0 & 0 & 0 & 0 \\ 0 & 0 & 0.8780 & 0 & 0 & 0 \\ 0 & 0.0734 & -0.0254 & 0.9908 & -0.1717 & 0 \\ 0 & 0.0018 & 0.1367 & 0.0498 & 0.9957 & 0 \\ 0.0496 & 0.0256 & 0.0044 & -0.0001 & -0.0030 & 1 \end{bmatrix}, \quad d^4 = \begin{bmatrix} 0.3199 \\ -0.8895 \\ 189.7399 \end{bmatrix},$$

$$A^5 = \begin{bmatrix} 0.9837 & -0.0079 & 0.1774 & 0 & 0.0013 & 0 \\ 0 & 0.9657 & 0 & 0 & 0 & 0 \\ 0 & 0 & 0.8780 & 0 & 0 & 0 \\ 0 & 0.1126 & -0.0254 & 0.9908 & -0.1712 & 0 \\ 0 & 0.0028 & 0.1367 & 0.0498 & 0.9957 & 0 \\ 0.0496 & 0.0239 & 0.0043 & -0.0001 & -0.0039 & 1 \end{bmatrix}, \quad d^5 = \begin{bmatrix} 0.4211 \\ -0.8501 \\ 249.3018 \end{bmatrix},$$

$$B^1 = \begin{bmatrix} 0.0047 & -0.0003 \\ 0 & 0.0491 \\ 0.0469 & 0 \\ -0.0005 & 0.0004 \\ 0.0035 & 0 \\ 0.0001 & 0.0007 \end{bmatrix}, \quad B^2 = \begin{bmatrix} 0.0047 & -0.0003 \\ 0 & 0.0491 \\ 0.0469 & 0 \\ -0.0005 & 0.0005 \\ 0.0035 & 0 \\ 0.0001 & 0.0007 \end{bmatrix},$$

$$B^3 = \begin{bmatrix} 0.0047 & -0.0002 \\ 0 & 0.0491 \\ 0.0469 & 0 \\ -0.0005 & 0.0010 \\ 0.0035 & 0 \\ 0.0001 & 0.0007 \end{bmatrix}, \quad B^4 = \begin{bmatrix} 0.0046 & -0.0002 \\ 0 & 0.0491 \\ 0.0469 & 0 \\ -0.0005 & 0.0018 \\ 0.0035 & 0 \\ 0.0001 & 0.0006 \end{bmatrix},$$

$$B^5 = \begin{bmatrix} 0.0045 & -0.0002 \\ 0 & 0.0491 \\ 0.0469 & 0 \\ -0.0005 & 0.0028 \\ 0.0035 & 0 \\ 0.0001 & 0.0006 \end{bmatrix}, \quad L^i = B^i, \ \forall_{i \in \{1, \ldots, 5\}},$$

$$C^1 = \begin{bmatrix} 0 & 0 & 0\ 0\ 0\ 1 \\ 0 & 0 & 0\ 0\ 1\ 0 \\ 0 & 896.2360 & 0\ 0\ 0\ 0 \end{bmatrix}, \quad C^2 = \begin{bmatrix} 0 & 0 & 0\ 0\ 0\ 1 \\ 0 & 0 & 0\ 0\ 1\ 0 \\ 0 & 894.1477 & 0\ 0\ 0\ 0 \end{bmatrix},$$

$$C^3 = \begin{bmatrix} 0 & 0 & 0\ 0\ 0\ 1 \\ 0 & 0 & 0\ 0\ 1\ 0 \\ 0 & 879.0008 & 0\ 0\ 0\ 0 \end{bmatrix}, \quad C^4 = \begin{bmatrix} 0 & 0 & 0\ 0\ 0\ 1 \\ 0 & 0 & 0\ 0\ 1\ 0 \\ 0 & 851.0135 & 0\ 0\ 0\ 0 \end{bmatrix},$$

$$C^5 = \begin{bmatrix} 0 & 0 & 0\ 0\ 0\ 1 \\ 0 & 0 & 0\ 0\ 1\ 0 \\ 0 & 810.7468 & 0\ 0\ 0\ 0 \end{bmatrix},$$

$$u^1 = \begin{bmatrix} 0 \\ 0 \end{bmatrix}, u^2 = \begin{bmatrix} 0 \\ 0.05 \end{bmatrix}, u^3 = \begin{bmatrix} 0 \\ 0.1 \end{bmatrix}, u^4 = \begin{bmatrix} 0 \\ 0.15 \end{bmatrix}, u^5 = \begin{bmatrix} 0 \\ 0.2 \end{bmatrix},$$

The reference input is defined by

$$u_h = u_{k,1} = 0,$$

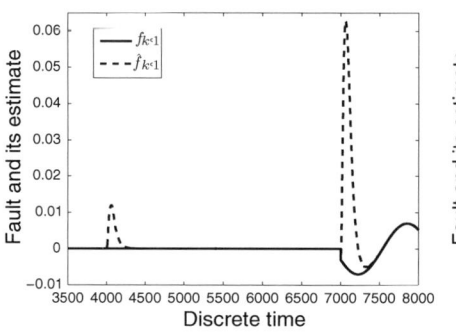 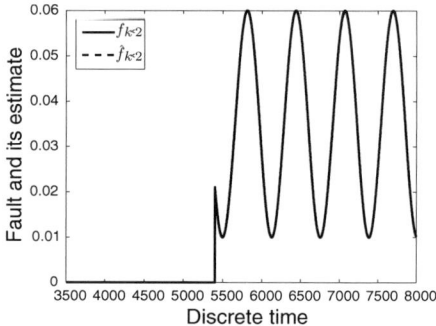

Fig. 6.15 Faults and their estimates

$$
\boldsymbol{u}_v = \boldsymbol{u}_{k,2} =
\begin{cases}
0, & k < 4000 \\
0.05, & 4000 \leqslant k < 7000 \\
0.10, & k \geqslant 7000
\end{cases}.
$$

The actuator fault scenarios, i.e., a decrease in the performance of the two rotors, are described as follows:

$$
\boldsymbol{f}_{k,1} =
\begin{cases}
0, & k < 7000, \\
0.007\sin(0.005k), & k \geqslant 7000,
\end{cases}
$$

$$
\boldsymbol{f}_{k,2} =
\begin{cases}
0, & k < 5400, \\
0.035 + 0.025\sin(0.01k), & k \geqslant 5400.
\end{cases}
$$

The regulator for the FTC system was designed with the following parameters $\gamma = 0.05$ and the weighting matrices, based on Bryson's rule [40]:

$$
\boldsymbol{Q}_R = \frac{1}{\gamma^2}\begin{bmatrix}1 & 0\\0 & 1\end{bmatrix},\quad
\boldsymbol{R}_R = \frac{1}{0.04}\begin{bmatrix}1 & 0\\0 & 1\end{bmatrix}.
$$

Due to the high transients of the systems in the initial phase, the FTC system was not enabled until the 4,000th iteration, but the control input was also in the neutral state (i.e., $\boldsymbol{u} = [0, 0]^T$) during that time period.

Figures 6.17 and 6.18 present the results achieved for the proposed FTC strategy (with $\alpha_k = \boldsymbol{I}$). As a result, Fig. 6.15 clearly shows that the faults can be estimated with a very high accuracy (especially the second fault estimator). The estimator for the first fault presents some deviations from the nominal value due to the abrupt changes in the reference input and modelling errors as a consequence of the high non-linearity of the system (its high cross-coupling between the main rotor and the tail rotor). But after about 250 samples, in both the faultless and the faulty state, estimation achieves very high accuracy, even though the fault is time-varying.

From Fig. 6.16, it can be observed that $\boldsymbol{u}_{f,k}$ is close to $\boldsymbol{u}_k$ until the occurrence of the fault $\boldsymbol{f}_k$. After that time, the control strategy $\boldsymbol{u}_{f,k}$ is changed. The only

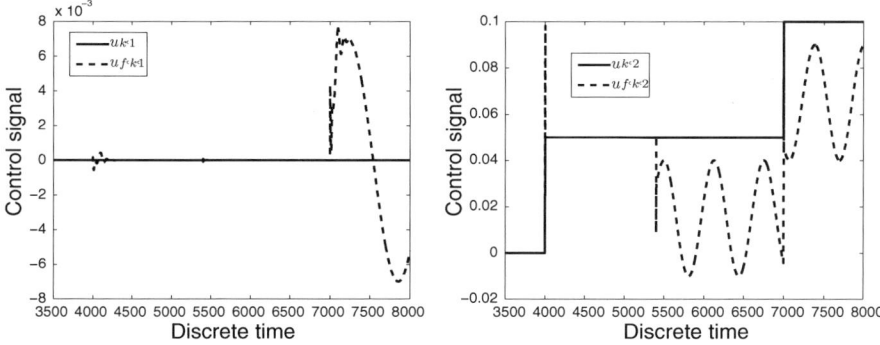

Fig. 6.16 Trajectories of $\boldsymbol{u}_k$ and $\boldsymbol{u}_{f,k}$

deviations from the expected behaviour can be seen in the cases where the faults were overestimated, due to the reasons stated above.

The final conclusion is that FTC stabilises the system with high performance (Figs. 6.17 and 6.18). Indeed, even in the presence of the faults the original trajectories are unchanged. This is because of the proposed control strategy, for which $\boldsymbol{x}_{f,k}$ converges to $\boldsymbol{x}_k$ and consequently $\boldsymbol{z}_k = \boldsymbol{y}_{f,k} - \boldsymbol{y}_k$ converges to zero, whereas the system without the FTC control significantly deviates from the original trajectories. Especially interesting is the fact that sometimes even a small uncompensated fault, as the one of the tail motor, can lead to catastrophic failure, which can be seen by the trajectories of the pitch angle of the beam α_h.

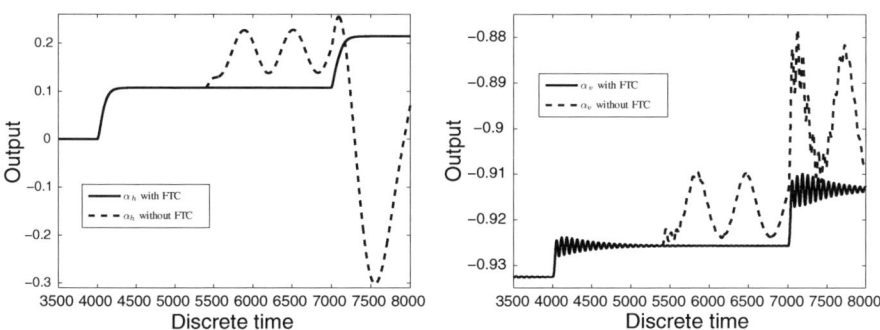

Fig. 6.17 Outputs of the system $\boldsymbol{y}_{f,k}$ with with and without FTC

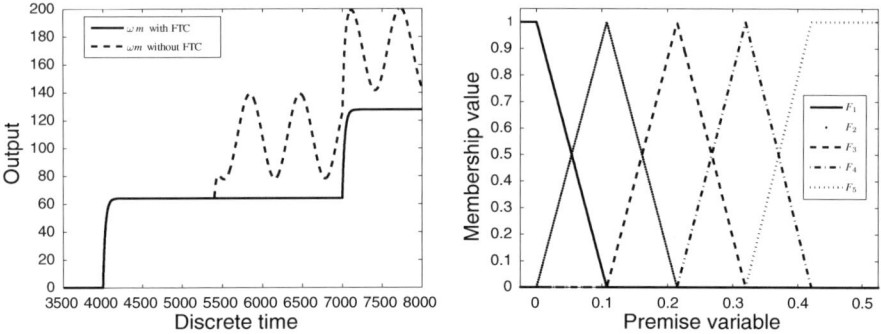

Fig. 6.18 Outputs of the system $\mathbf{y}_{f,k}$ with and without FTC as well as fuzzy sets used in the Takagi–Sugeno model

6.7 Concluding Remarks

In this chapter, an active FTC strategy was proposed, which enables on-line reconfiguration of control after the occurrence of sensor and actuator faults. The approach was developed in the context of Takagi–Sugeno fuzzy systems. The key contribution of the proposed approach is an integrated FTC design procedure of fault identification and fault-tolerant control schemes. The procedure also allows including input constraints into the FTC system. The FTC controller is implemented as a state feedback controller. This controller is designed in such a way that it can stabilize the faulty plant using the Lyapunov theory and LMIs. The designed controller is called a virtual actuator, because it feeds the faulty actuator and produces the intended effects on the output of the plant by using an appropriate compensation. Also a design procedure for the regulator for TS fuzzy systems was shown, which enables minimising the objective cost function. The chapter presented also a number of illustrative examples, which exhibit the performance of the proposed approaches. In particular, it is shown how to design the TS model of a tunnel furnace. The procedure starts with an appropriate selection of experimental design, i.e., the shape and frequency of the system input. Subsequently, the data-driven strategy is used to derive the TS model. The obtained model is then used to design the virtual sensor, which makes it possible to implement the FTC strategy with respect to sensor faults. Another example concerns the celebrated benchmark, i.e., a three-tank system. In this case, the virtual fuzzy actuator is examined. The same approach is tested on a twin-rotor system, which mimics the helicopter. The systems being considered possesses completely different properties but in both cases the proposed FTC strategy proved to be an efficient tool.

References

1. L.A. Zadeh, Fuzzy Sets. Inf. Control. **8**, 338–353 (1965)
2. J.M. Mendel, Fuzzy sets for words: a new beginning. *International Conference on Fuzzy Systems* (St. Louis, 2003), pp. 37–42
3. L. Dziekan, A. Marciniak, A. Obuchowicz, Segmentation of colour cytological images using type-2 fuzzy sets, in *Fault Diagnosis and Fault Tolerant Control, Challenging Problems of Science–Theory and Applications: Automatic Control and Robotics*, ed. by J. Korbicz, K. Patan, M. Kowal (Academic Publishing House EXIT, Warsaw, 2007), pp. 263–270
4. J.M. Mendel, Fuzzy logic systems for engineering: a tutorial. Proc. IEEE **83**, 345–377 (1995)
5. L. Dziekan, M. Witczak, J. Korbicz, Active fault-tolerant control design for Takagi-Sugeno fuzzy systems. Bull. Polish Acad. Sci. Tech. Sci. **59**(1), 93–102 (2011)
6. T. Takagi, M. Sugeno, Fuzzy identification of systems and its application to modeling and control. IEEE Trans. Syst. Man Cybern. **15**(1), 116–132 (1985)
7. Kazuo Tanaka, Hua O. Wang, *Fuzzy Control Systems Design and Analysis: A Linear Matrix Inequality Approach* (Wiley-Interscience, New York, 2001)
8. Piotr Tatjewski, *Advanced Control of Industrial Processes: Structures and Algorithms Advances in Industrial Control* (Springer, London, 2007)
9. J. Korbicz, J. Kościelny, Z. Kowalczuk, W. Cholewa (eds.), *Fault Diagnosis. Models, Artificial Intelligence, Applications* (Springer, Berlin, 2004)
10. T.M. Guerra, A. Kruszewski, J. Lauber, Discrete Takagi-Sugeno models for control: where are we? Annu. Rev. Control **33**(1), 37–47 (2009)
11. J.L. Castro, M. Delgado, Fuzzy systems with defuzzification are universal approximators. IEEE Trans. Syst. Man Cybern. **26**(1), 149–152 (1996)
12. K. Gasso, G. Mourot, J. Ragot, Structure identification in multiple model representation: elimination and merging of local models, *40th IEEE Conference on Decision and Control* (Orlando, 2001), pp. 2992–2997
13. M. Margaliot, G. Langholz, A new approach to fuzzy modeling and control of discrete-time systems. IEEE Trans. Fuzzy Syst. **11**(4), 486–494 (2003)
14. Y. Lu, Y. Arkun, Quasi-min-max MPC algorithms for LPV systems. Automatica **36**(4), 527–540 (2000)
15. M. Sugeno, G. Kang, Structure identification of fuzzy model. Fuzzy Sets Syst. **26**(1), 15–33 (1988)
16. T.A. Johansen, R. Shorten, R. Murray-Smith, On the interpretation and identification of dynamic Takagi-Sugeno fuzzy models. IEEE Trans. Fuzzy Syst. **8**(3), 297–313 (2000)
17. J. Espinosa, J. Vandewalle, V. Wertz, *Fuzzy Logic, Identification and Predictive Control (Advances in Industrial Control)* (Springer, Berlin, 2004)
18. Ch. Li, J. Zhou, X. Xiang, Q. Li, X. An, T-S fuzzy model identification based on a novel fuzzy c-regression model clustering algorithm. Eng. Appl. Artif. Intell. **22**(4–5), 646–653 (2009)
19. H.O. Wang, K. Tanaka, M.F. Griffin, An approach to fuzzy control of nonlinear systems: stability and design issues. IEEE Trans. Fuzzy Syst. **4**(1), 13–23 (1996)
20. S. Hui, S.H. Zak, Observer design for systems with unknown input. Int. J. Appl. Math. Comput. Sci. **15**(4), 431–446 (2005)
21. M. Witczak, *Modelling and Estimation Strategies for Fault Diagnosis of Non-linear Systems* (Springer, Berlin, 2007)
22. Q. Rong, G.W. Irwin, LMI-Based controller design for discrete polytopic LPV systems, *European Control Conference* (United Kingdom, Cambridge, 2003), pp. 1–6
23. S. Boyd, L.E. Ghaoui, E. Feron, V. Balakrishnan, *Linear Matrix Inequalities in System and Control Theory, volume 15 of Studies in Applied Mathematics* (SIAM, Philadelphia, 1994)
24. P.M. Frank, Fault diagnosis in dynamic system via state estimation-a survey, in *System Fault Diagnostics, Reliability, Related Knowledge-based Approaches*, vol. 1, ed. by S. Tzafestas, M. Singh, G. Schmidt (eds.) (D. Reidel Press, Dordrecht, 1987), pp. 35–98
25. R.J. Patton, P.M. Frank, R.N. Clark, *Fault Diagnosis in Dynamic Systems: Theory and Applications* (Prentice-Hall, Englewood Cliffs, 1989)

26. M. Blanke, M. Kinnaert, J. Lunze, M. Staroswiecki, *Diagnosis and Fault-Tolerant Control* (Springer, New York, 2003)
27. J. Chen, R.J. Patton, *Robust Model-based Fault Diagnosis for Dynamic Systems* (Kluwer Academic Publishers, London, 1999)
28. GE Fanuc Automation. PACSystems CPU Reference Manual, GFK-2222F (2005)
29. LUMEL S.A. User Guide, Single-Phase Semiconductor Relay RP6, 1 2004. (in Polish).
30. GE Fanuc Automation. PACSystems RX3i System Manual, GFK-2314C (2005)
31. GE Fanuc Automation. Hardware User's Guide, 15" QuickPanel View & QuickPanel, Control GFK-2402 (2005)
32. W. Favoreel, B. De Moor, P. Van Overschee, Subspace state space system identification for industrial processes. J. Process Control **10**(2–3), 149–155 (2000)
33. L. Ljung, *System Identification—Theory for the User*, 2nd edn. (Prentice Hall, Upper Saddle River, 1999)
34. T.F. Coleman, Y. Li, An interior, trust region approach for nonlinear minimization subject to bounds. SIAM J. Optim. **6**(2), 418–445 (1996)
35. A. Zolghadri, D. Henry, M. Monsion, Design of nonlinear observers for fault diagnosis. A case study. Control Eng. Pract. **4**(11), 1535–1544 (1996)
36. A. Akhenak, M. Chadli, D. Maquin, J. Ragot, State estimation via multiple observer: The three tank system, *5th IFAC Symposium Fault Detection Supervision and Safety of Technical Processes (SAFEPROCESS)* (USA, Washington DC, 2003), pp. 1227–1232
37. Feedback Instruments Limited. Twin Rotor MIMO System Advanced Teaching Manual 1 (Crowborough, 1998)
38. A. Rahideh, M.H. Shaheed, Mathematical dynamic modelling of a twin-rotor multiple input-multiple output system. Inst. Mech. Eng. Part I J. Syst. Control Eng. **227**, 89–101 (2007)
39. S. Montes de Oca, V. Puig, M. Witczak, J. Quevedo, Fault-tolerant control of a two-degree of freedom helicopter using LPV techniques, *16th Mediterranean Conference on Control and Automation* (Ajaccio, 2008), pp. 1204–1209
40. G.F. Franklin, J.D. Powell, A. Emami-Naeini, *Feedback Control of Dynamic Systems*, 4th edn. (Prentice Hall, Upper Saddle River, 2002)

Chapter 7
Conclusions and Future Research Directions

From the point of view of engineering, it is clear that providing a reliable control strategy that is able to tolerate potential faults is an essential issue in modern control design, particularly as far as the control of complex industrial systems is considered.

There is no doubt that such a challenging task can only be realised with suitably integrated fault diagnosis and control schemes. The only way to realise efficient integration is to obtain as much information about the faults as possible while taking into account any potential imprecision of the implemented fault diagnosis.

Unfortunately, most systems present in our reality exhibit non-linear behaviour, which makes it impossible to use the well-developed techniques for linear systems. If it is assumed that the system is linear, which is not true in general, and even if robust techniques for linear systems are used (e.g., unknown input observers, linear state-feedback controllers, etc.), it is clear that such an approximation may lead to unreliable performance of fault diagnosis and fault-tolerant control. Indeed, due to the imprecision of the system description, early indication of faults which are developing is rather impossible. Such a situation increases the probability of the occurrence of failures, which can be extremely serious in terms of economic losses, environmental impact, or even human mortality.

Indeed, robust techniques are able to tolerate a certain degree of model uncertainty. In other words, they are not robust to everything, i.e., are robust to an arbitrary degree of model uncertainty. This real world development pressure creates the need for new techniques which are able to tackle integrated fault diagnosis and fault-tolerant control for non-linear systems. As discussed in introduction, in spite of the fact that the problem has been attacked from various angles by many authors and a number of relevant results have already been reported in the literature, there is no general framework which can be simply and conveniently applied to maintain such an integration.

As was underlined in Chap. 2, observers are immensely popular as residual generators for fault detection (and, consequently, for fault isolation) of both linear and non-linear dynamic systems. Their popularity lies in the fact that they can also be

M. Witczak, *Fault Diagnosis and Fault-Tolerant Control Strategies for Non-Linear Systems*, 221
Lecture Notes in Electrical Engineering 266, DOI: 10.1007/978-3-319-03014-2_7,
© Springer International Publishing Switzerland 2014

employed for control purposes. This natural property makes them a perfect candidate for realising the required integration procedure.

There are, of course, many different observers (or filters in the stochastic case) which can be applied to non-linear, and especially non-linear deterministic systems. Logically, the number of "real world" applications (not only simulated examples) should proliferate, yet this is not the case. The main reason is that the design complexity of most observers for non-linear systems does not encourage engineers to apply them in practice. Moreover, their design procedures are usually presented with academic artificial examples, which are useless for engineers. Another reason is that the applicability of robust observers, such as the celebrated unknown input observer, is limited due to the lack of an appropriate description of model uncertainty.

The above discussion clearly justifies the need for simpler observer structures, which can be obtained by solving the following problems:

Problem 1   Improvement of convergence of linearisation-based observers.

Problem 2   Design of data-driven methods for determining the model uncertainty, i.e., derivation of an unknown input distribution matrix.

Problem 3   Design of filters that are able to switch (or mix) unknown input distribution matrices according to the operating conditions.

As was mentioned in Chaps. 3 and 6, challenging design problems arise regularly in modern fault diagnosis systems and fault-tolerant control. Unfortunately, the classic analytical techniques often cannot provide acceptable solutions to such difficult tasks. If this is the case, one possible approach is to use soft computing-based fault diagnosis approaches, which can be divided into three categories:

- neural networks,
- fuzzy logic-based techniques,
- evolutionary algorithms.

Apart from the unquestionable appeal of soft computing approaches, there is a number of design issues that can be described by

Problem 4   Development of robust neural network-based fault diagnosis.

Problem 5   Integration of analytical and soft computing fault-tolerant control.

The issue of integrated fault diagnosis and fault-tolerant control raises other problems:

Problem 6   Design of a general framework for integrated fault diagnosis and fault-tolerant control.

Problem 7   Extension of the developed framework for non-linear systems described by analytical models.

Problem 8   Minimisation of the uncertainty effect (modelling uncertainty, disturbances, noise) on the performance of integrated fault diagnosis and fault-tolerant control.

Problem 9   Extension of the developed framework for non-linear systems described by soft computing-based models.

Although partial solutions to *Problems* 1–9 are scattered over many papers and a number of book chapters, there is no work that summarises all of these results in a unified framework.

Thus, one original objective of this book was to present selected approaches for solving the challenging *Problems* 1–9 in a unified framework.

Other objectives, perceived as solutions to *Problems* 1–9, are presented in the form of a concise summary of the contributions provided by this book to the state-of-the-art of modern model-based fault diagnosis for non-linear systems:

Solutions to Problem 1

- A general scheme of the EUIO is proposed that is based on the second-order extended Kalman filter (cf. Sect. 2.3). The convergence condition is also provided and employed to improve the convergence of the described approach.
- It was also empirically verified that the performance of the EUIO can be suitably extended, which recommends its practical application.

Solution to Problem 2

- A design strategy for developing the unknown input distribution matrix is proposed (cf. Sect. 2.5). This data-driven optimisation approach is based on the approach that prevents the unappealing effect of fault decoupling (cf. Sect. 2.2). This means that the determined unknown input distribution matrix will not cause this unappealing phenomenon.

Solutions to Problem 3

- The unknown input filter, which is based on the unscented Kalman filter, is proposed in Sect. 2.4. The approach can be applied for non-linear stochastic systems.
- The interactive multiple model approach along with the proposed unknown input filter is employed for settling the problem of switching (or mixing) unknown input distribution matrices according to the operating conditions.

Solutions to Problem 4

- Robust neural network-based fault diagnosis strategies are revisited in Chap. 3. The general framework presented in [1] is suitably extended by providing an alternative fault detection approach, which is called the backward test (cf. Sect. 3.2.5).
- It was empirically proven that the backward test constitutes an appealing alternative to the well-known forward one.

Solution to Problem 5

- Chapter 6 provides an integrated fault-tolerant control scheme for the non-linear systems that can be described with Takagi–Sugeno models. The proposed approach is based on analytical techniques and is carefully described in Chap. 4.

Solutions to Problem 6

- A general framework for integrated fault diagnosis and fault-tolerant control is proposed in Chap. 4.

- The proposed approach is based on the observer-based fault identification scheme that is suitably integrated with state-feedback control. It is worth underlining that the appealing property of the proposed framework is that it takes into account the imprecision related to the fault diagnosis procedure.

Solution to Problem 7

- The general fault-tolerant control framework proposed in Sect. 4.1.4, is extended to Lipschitz non-linear discrete-time systems and carefully described in Chap. 4.

Solutions to Problem 8

- The general fault-tolerant control framework proposed in Chap. 4 is extended in Chap. 5 to handle the robustness issue.
- Using the $\mathcal{H}_\infty$ approach, the uncertainties present in the state equation are tackled in Sect. 5.1.
- Applying the $\mathcal{H}_\infty$ strategy, the uncertainties present in both the state and output equations are tackled in Sect. 5.2.
- Using the $\mathcal{H}_\infty$ approach, non-linear systems described by quasi-LPV models with the uncertainties in the state and output equations are tackled in Sect. 5.3.

Solutions to Problem 9

- The framework proposed in Sect. 4.1.4 was suitably extended to the systems that can be described with the Takagi–Sugeno fuzzy models (cf. Chap. 6).
- The fault-tolerant control scheme was also extended to the case of the sensor faults (cf. Chap. 6).

The book also presents a number of practical implementations of the proposed approaches, which can be summarised as follows:

- State estimation and sensor, actuator fault diagnosis of a two-phase induction motor;
- State estimation and sensor, actuator fault diagnosis of a two-tank system;
- Experimental design for a neural model of a valve actuator;
- MLP-model design of a valve actuator;
- MLP-based robust fault detection of a valve actuator;
- GMDH neural network-based model design of a valve actuator;
- GMDH neural network-based robust fault detection of a valve actuator with forward and backward tests;
- Fault-tolerant control of a twin-rotor system using analytical techniques;
- Fault-tolerant control of a twin-rotor system using fuzzy logic;
- Fuzzy model design of a tunnel furnace;
- Fault-tolerant control of a tunel furnace;
- Fault-tolerant control of a three-tank system.

The advantage of the general framework presented in this monograph is the fact that it is independent of a particular form of the system being diagnosed. Indeed, when the non-linear state-space model is available, then effective observer-based approaches can be employed. If this is not the case, then one can design such models with the proposed Takagi–Sugeno approach. An alternative solution is to use the proposed robust neural network-based techniques.

Irrespective of the above advantage, there still remain open problems regarding some important design issues. What follows is a discussion of the areas proposed for further investigations.

Handling constraints in the fault-tolerant control There is no doubt that all systems have physical constraints which limit their performance. These constraints may change during fault occurrence. For example, when an actuator fault appears in systems, then the control limits must be suitably modified.

Development of integrated fault-tolerant control with neural networks It was proven that neural networks can be efficiently used for robust faul diagnosis. Thus, a natural task seems to employ them for fault-tolerant control within a unified framework described in Chap. 4.

Relaxing the existence conditions of observer-based fault-tolerant control As has been mentioned, many non-linear systems satisfy Lipschitz conditions. On the other hand, the Lipschitz constant for such systems may have relatively large values. This can make the usage of the observer design procedures described in Chap. 5 impossible. Thus, the development of less conservative transformation techniques constitutes one of the future research directions.

Reference

1. M. Witczak, *Modelling and Estimation Strategies for Fault Diagnosis of Non-linear Systems* (Springer-Verlag, Berlin, 2007)

Index

M. Witczak, *Fault Diagnosis and Fault-Tolerant Control Strategies for Non-Linear Systems*, 227
Lecture Notes in Electrical Engineering 266, DOI: 10.1007/978-3-319-03014-2,
© Springer International Publishing Switzerland 2014